Heredity, Development and Evolution

Christine Birkett

B.Sc., Certificate in Education, A. Inst. Biol;
formerly Biology department, Weymouth Grammar School,
Lecturer, Stoke-on-Trent Cauldon College of Further Education

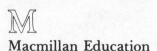

Macmillan Education

To John

First published 1979
Reprinted 1982, 1983, 1984, 1985, 1986, 1987

Published by
MACMILLAN EDUCATION LTD
Houndmills, Basingstoke, Hampshire RG21 2XS
and London
Companies and representatives
throughout the world

Printed in Hong Kong

ISBN 0-333-24192-4

Contents

FOUNDATIONS OF BIOLOGY
General editor L. M. J. Kramer
A major advanced biology course
    for schools and colleges

Books in the series are:
The diversity of life
The cell concept
Heredity, development and evolution
Metabolism, movement and control
Man and the ecosystem

Preface

Foundations of biology aims to provide a complete pre-university course in biological science. Accordingly, the work is covered in a few handy volumes, not in a single bulky one or numerous monographs. The questions at the ends of the chapters are to test the comprehension of the material covered in the chapters and their contents are not necessarily similar to those set in biological examinations which often require knowledge in several branches of biology if they are to be answered properly. Suggestions are provided for further reading.

The course consists of five books written by experienced teachers with special knowledge of biological science, who believe through their experience that fresh approaches to teaching biology are desirable at pre-university level. The books in the series are:

The diversity of life
The cell concept
Heredity, development and evolution
Metabolism, movement and control
Man and the ecosystem.

Biologists will realise the difficulty of sub-dividing the course into a number of books and opinions will undoubtedly differ on how it should best be done. One difficulty is that a number of topics are based upon knowledge of others, so that if each book is to be helpful some overlap must occur with others in the series. In fact, the necessity for overlap has proved to be relatively small and where it occurs the treatment of topics is consistent from one book to another.

It is wise to remember that no branch of science is more 'fundamental' than any other, so no suggestion has been made that the books need to be studied in a given order. Teachers will be free to use them in any sequence or combination which suits their own courses.

All the authors concerned with the series have felt keenly the inadequacy of purely descriptive biology in giving insight into the basis of science today. It has been necessary therefore for them to introduce some mathematics, physics and organic chemistry to which biology is so closely related. The names of chemical compounds are accompanied by their new names under the IUPAC rules and in *The cell concept* there is an introduction to the new uses which seem difficult at first but which are in fact logical and easy to follow once the principles have been grasped.

L. M. J. KRAMER *General Editor*

Acknowledgements

The author would like to thank Dr J. T. Williams of the Food and Agriculture Organization of the UNO for providing material used in chapter 6.

The author and publishers wish to thank the following who have kindly given permission for the use of the copyright material:-

Artemis Press Limited for an illustration from *Understanding the Earth* edited by I. G. Grass, P. J. Smith and R. L. C. Wilson. Associated Book Publishers Limited and Professor E. B. Ford for an amalgamation of figures taken from *Mendelism and Evolution 7/E*, published by Methuen and Company Limited and *Ecological Genetics*, published by Chapman and Hall Limited; and drawings based on a photograph (Plate 1) from *Ecological Genetics*. The Associated Examining Board (for the General Certificate of Education) for questions from *Advanced Level Biology* Paper 2 & 3, June 1975, and Paper 3, November 1975.
Cambridge University Press for a table from *Basic Biology Course*, Book 9 by M. A. Tribe, I. Fallan, M. R. Grout, and R. K. Snook, University of Sussex. Wm. Collins Sons & Co. Limited for figures from *Pollination of Flowers* by Proctor and Yeo, in the New Naturalist Series. W. H. Freeman and Company for illustrations from 'Symbiosis and Evolution' by Lynn Margulis in *Scientific American*, August 1971 and 'Prenatal Diagnosis of Genetic Disease' by T. Friedmann in *Scientific American*, November 1971. Heinemann Educational Books Limited for a figure from *Plant Growth* by M. Black and J. Edelman in *Foundations of Biology*. Longman Group Limited for a table from Nuffield Advanced Biology (1970) *Organisms and Populations*, *A Laboratory Guide*. Macmillan Publishing Co. Inc., for figures from *Genetics* by Monroe W. Strickberger, copyright © 1968, 1976, Monroe W. Strickberger. W. B. Saunders Company and Professor B. I. Balinsky for a figure from *Introduction to Embryology* © 1965, W. B. Saunders Company, Philadelphia. Professor J. J. Murray, Jnr., for a figure from *Genetic Diversity and Natural Selection* (based on a paper by Dr Littlejohn in *Evolution*, 1965, Volume 19).

The author and publishers wish to acknowledge the following photograph sources:

Heather Angel p. 137. Barnabys Pictures Library p. 63 bottom left. Gene Cox p. 135. Patrick Echlin, Botany School, Cambridge p. 112. John Innes Institute p. 161. Dr Kettlewell, Department of Zoology, Oxford p. 92. Dr G. F. Leedale p. 63 top and bottom right. Dr M. C. F. Proctor pp. 114, 115. C. James Webb pp. 94, 133. Zoological Society of London p. 182. Taken from the publication Looking at Chromosomes, McLeish & Snoad p. 35. Taken from the publication – Webster – Introduction to Fungi – Cambridge University Press p. 49.

The publishers have made every effort to trace copyright holders, but if they have inadvertently overlooked any they will be pleased to make the necessary arrangements at the first opportunity.

1 Evolution

One of the most striking facts emerging from a study of natural history is the variety of life; another, paradoxically, is its underlying unity. While there are about two million kinds of *organisms* (living creatures) alive today, each of them can be grouped with others which it resembles. Each of the three-quarters of a million insects, for example, has a body divided into a head, three-part thorax and abdomen. Almost every backboned animal, or *vertebrate*, has an internal rigid skeleton with four limbs. There are, moreover, different degrees of similarity and difference between living things, which suggests that organisms are variations on a few basic 'themes'. Within each theme are other groups with more characteristics in common, and sub-groups within these yet more similar; thus all beetles resemble one another more closely than do all insects, and all ladybird beetles are more alike still. This hierarchy of variation has stimulated men of all times to classify living forms: we have evidence that the ancient Greeks did so. Indeed, the Greek naturalist and philosopher, Aristotle (384–322 B.C.), aware that variations on a theme are seldom generated independently and separately, put forward the view that there was a gradual change in living forms from plants to animals to man. This was the first recorded hypothesis that *organic evolution* had occurred, that is that living organisms are descended from earlier, different ones.

There is an alternative viewpoint to that expressed by the idea of organic evolution. It is possible that each kind of organism alive today was created separately and has not altered since its creation; that there is no relationship by descent between living forms. This hypothesis is called 'Special Creation'. By the eighteenth and nineteenth centuries, the idea that each *species* (a group of organisms which does not breed with any other group under natural conditions) had been specially created was much more widely accepted than the view that each had evolved. The Bible, after all, declared that each living thing was 'brought forth after his kind' and the usual interpretation of the statement was a literal one: that a 'kind' or species was unchangeable.

In 1794, a British physician and naturalist, Erasmus Darwin, published a great work called *Zoonomia*, in which he proposed that all animals had arisen from one simple living creature by becoming better adapted to their environment each generation. At about the same time, a French professor of zoology, Jean-Baptiste Lamarck, suggested a mechanism by which such changes might come about. Lamarck believed that animals and plants respond to their environments by becoming better adapted to them. Bears in cold places grow thicker fur, for example, and plants in shady places develop longer stems and so reach the light. Once such a small 'improvement' is acquired, Lamarck maintained that it would be inherited by its possessor's offspring. In the offspring, the changing character would become even more pronounced under the environmental influence, until much later descendants were quite different from the original form. This theory, known as 'Lamarckism' or 'the inheritance of acquired characteristics' was a first bold attempt to suggest a mechanism by which organic evolution could occur, adapting living creatures more and more perfectly to their surroundings.

Lamarckism is not borne out by experiment, however. There is no evidence that changes brought about during the course of an organism's life are inherited by its offspring: indeed all evidence is to the contrary. A blacksmith's son will not have particularly powerful arms unless he subjects them to the same muscular effort as did his father. Despite failing to describe a plausible *mechanism* for evolution, Lamarck was successful in drawing attention to the *fact* of evolution at a time when the majority believed species to be unchanged and unchanging.

CHARLES DARWIN

The man destined to make the greatest, lasting impact on evolutionary theory, Charles Darwin, grandson of Erasmus, was born in 1809. After abandoning a medical career at Edinburgh, Darwin spent three years at Cambridge studying theology. He worked sufficiently hard to obtain a pass degree, but his main interests lay in geology and botany. The professors of these two subjects, Adam Sedgwick and J. S. Henslow, must have recognised Darwin's potential, for they encouraged him in the ways of science. After graduating, his heart not in the Church, Darwin heard from Henslow that Captain Fitzroy was looking for a naturalist to accompany a world survey using his ship, H.M.S. *Beagle*. Darwin was appointed to the post and set sail on 27 December 1831.

The voyage of the 'Beagle'

The *Beagle* was at sea for five years, during which time Darwin avidly collected and studied plant, animal and rock specimens. His reading material included volume 1 of Lyell's *Principles of Geology*. Lyell understood that the rocks found on the earth arose by slow changes in the earth's surface, and that the forces producing such changes still operated. Lyell's principle that 'the present is the key to the past' appealed to Darwin. Aided by Lyell's book, Darwin concluded that certain coastal fossil-bearing rocks, towering high above the water, had once formed part of the sea-bed.

In the rich South American fossil-beds, Darwin discovered many species of extinct animals and he noted a similarity between the fossilised and living armadillos, tapirs and anteaters. This observation of resemblance in fundamental design but not in detail between fossil and living creatures was crucial to Darwin's later theories.

The Galápagos Islands

In 1835, the *Beagle* spent five weeks in the Galápagos Islands, an archipelago in the Pacific Ocean. Darwin was impressed by the differences between the living creatures and those on the nearest mainland of Ecuador, 950 kilometres (600 miles) to the east. Marine lizards, peculiar to the Galápagos were common, and each major island had a race of giant tortoise distinct from that of every other island.

The birds of the archipelago were remarkable. Of twenty-six kinds which Darwin captured, twenty-five were *endemic* (peculiar) to the islands. They included thirteen species of finch, fundamentally alike in body-form and plumage, but differing in numerous respects, notably in beak size and shape. Darwin wrote '. . . one might really fancy that from an original paucity of birds in this archipelago, one species had been taken and modified for different ends.' He was suggesting that these birds had all evolved from modifications of one ancestor yet, at the time of writing, he had not completely shed the belief of the times that species were immutable or unchangeable (Figure 1).

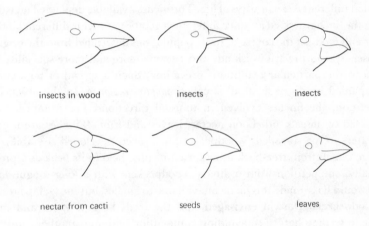

Figure 1   Some of the thirteen types of Galápagos finch showing adaptations to different diets

In October 1836, the *Beagle* returned to Falmouth and Darwin turned his attention to sorting his notebooks and his vast collection.

Evolution by natural selection

After nine months occupied with writing journals and articles about his voyage, Darwin began an essay called 'The Transmutation of Species'. Within fifteen months he had formulated a theory of how new species were formed by modification (or 'transmutation') of older types and not by Special Creation.

Darwin was aware that sexual reproduction (involving two parents) gives rise to much more variable offspring than does asexual reproduction (involving only one parent). Moreover, much of this variation is inheritable. By itself, Darwin noted, a population does not alter with the passage of time merely because it is variable. Different groups of the population must be isolated from one another and subjected to different conditions of climate or food or predation before any consistent change appears between the groups. Now he was in a position to interpret his observations on the Galápagos finches.

The Galápagos Islands are volcanic, so they were uninhabited when they erupted from the ocean floor two million years ago. In Darwin's day, the islands had many forms of life which must have originated from plants and animals which managed to colonise them from the mainland. Darwin noted that, although the Galápagos finches were unique, they resembled the nearest mainland finches. Moreover, other types of bird were relatively scarce on the islands. Darwin reasoned that, long ago, one species of mainland finch was swept to the Galápagos Islands (perhaps by the South East Trade winds) in sufficient numbers to establish a breeding population. The fact that Darwin observed greater differences among the finches than among comparable groups of other birds suggests that the finches were the first to reach the Galápagos. Since there was no competition from other species of bird, opportunity existed for the finches to exploit different *niches* or ways of life. The niches available depended on which islands the finches arrived at, since different vegetations provide different kinds of shelter and food. In the course of time, populations descended from the original colonisers spread to various islands and became more and more specialised or *adapted* to the particular conditions prevailing. Such a spread of population, accompanied by diversification, is called *adaptive radiation*. In the absence of competition, the finches evolved in unusual directions. For instance, some finches fed on insects, others on nectar, leaves and fruit. (Finches are usually seed-eaters.) One so-called 'woodpecker-finch' even developed the ability to dislodge insects from tree-bark using a cactus spine held in its beak as a probe. Normally, competition from ordinary woodpeckers with a long tongue for a probe would have made this niche unavailable to finches, but the Galápagos has no woodpeckers. Darwin envisaged that the birds became more and more diverse in feeding habits, so avoiding competition with one another, until the most extreme types evolved into separate species. He thought that the relative isolation of different populations on various islands was crucial to this evolution; if left close together, inter-breeding would 'swamp' any tendency for groups to become differentiated.

Darwin believed that the resemblances he had noted between fossil and living species could be accounted for by the same reasoning: that there was a common ancestry from which they had diverged. In this case, however, some types survived whilst others became extinct.

This argument, by itself, did not explain how adaptations to a particular way of life might come about. Darwin had for a long time collected information on the alteration of domestic species of plants and animals by breeding according to human choice. He noted that the differences between varieties of specially-bred pigeons were greater than differences between certain other true species. (*Varieties* or *races* are capable of inter-breeding, whilst true species, with rare exceptions, cannot inter-breed.) Darwin perceived that the differences were brought about by choice or *selection* of breeding pairs: man noted a characteristic he wanted to encourage in his pigeons and bred two individuals in which that characteristic was most marked. This was repeated, generation after generation, until a variety widely different from the parent stock was produced.

The problem was to envisage what forces might be at work in nature 'selecting' the individuals best adapted for survival, just as man chooses those he prefers. In 1838, Darwin read 'An Essay on the Principle of Population' by the Rev. T. R. Malthus, a political economist. This explained how human populations tend to increase their numbers in a geometrical progression (1, 2, 4, 8, 16) while food supplies increase only by arithmetical progression (1, 2, 3, 4, 5). Many people in every generation therefore die as a direct or indirect result of competition for food. Darwin now saw the answer for which he was seeking. He reasoned that under natural conditions all organisms would be involved in a struggle for survival, since many more are born than can be sustained. He calculated the theoretical rate of population increase of elephants, these being one of the slowest breeding animals. He reckoned that in 750 years there would exist nineteen million living elephants descended from a single pair. This situation is avoided only because the majority of elephants never survive to breed.

Darwin reasoned that the small variations that occur between members of a species must sometimes benefit one individual more than another. Those which die or fail to multiply (resulting in a stable population) are not, therefore, a random sample. The successful ones which become parents are *naturally selected*, as fantail pigeons are artificially selected by man. In nature, the survivors are selected since they are better fitted than their competitors to their way of life because of the characteristics they possess. Darwin knew from his studies of pigeons that modifications of structure could be inherited, so here was the final link. The best adapted organisms survive and produce offspring which inherit their adaptations. In the course of time, favourable variations accumulate so that the whole nature of the species changes. This theory is called '*evolution by natural selection*'.

Here, then, Darwin had put together a logical theory of evolution which may be summarised briefly:

1   Living organisms have the potential to increase at an accelerating rate.

2   The numbers of a species tend to remain constant nevertheless.

3   Therefore there must be competition or 'a struggle for existence' between organisms for those finite resources that they all need.

4   All living organisms vary from one another.

5   Therefore in the competition, only the 'fittest,' or best-adapted, organisms will survive to breed.

6   The offspring of these 'naturally-selected' organisms will inherit many of their parents' favourable characteristics.

'The Origin of Species'

It seems natural for a man who believes he has discovered a fundamental truth to publish it as fast as possible. While Darwin had established all the major points of his theory by 1839, it was not for twenty years that the world read of it. He was a sick man for most of his life, it is true, and a cautious one who preferred to test and verify his ideas before committing them to print. None the less, twenty years is a long time, and during this period Darwin published on other topics as diverse as coral reefs and barnacles. Apparently, Darwin was reluctant to face the inevitable opposition to his theory; a religious man himself, he could see clearly that it seemed to deny a literal interpretation of the Bible.

Darwin's dilemma persisted until 1858 when he received a paper from Alfred Russel Wallace with a request for his comment. It must have been a shock. Darwin read a theory identical to his own: Wallace had come to the same conclusion whilst working in the Malay archipelago. Darwin probably regretted his delay and publication was now essential. As a tactful solution, papers by both men were read to the famous Linnaean Society in London. The papers made little impact, but Darwin immediately began to compress his findings into a book. In November 1859 *The Origin of Species* was published and this time the theory caused a sensation.

Some eminent British scientists, notably T. H. Huxley, the zoologist, Hooker the botanist and Lyell the geologist, supported Darwin's theory, but there was also a lot of opposition. In a famous debate at the Oxford meeting of the British Association in 1860 matters came to a head and Huxley convincingly argued Darwin's case. Biologists in other countries, also, were soon persuaded and within a dozen years the theory had gained worldwide acceptance.

Problems with the theory

The theory of evolution by natural selection was the most important scientific idea of the nineteenth century and it still stands today with minor modifications.

Darwin's chief difficulty was that he did not understand how organisms provided constantly varying forms each generation on which natural selection could act. It was generally accepted that inheritance is *blending*, that is, some substance is passed from each parent to the offspring in which the two contributions mix. This substance was supposed to be derived from every part of the parent's body and to accumulate eventually in the *gametes* (the eggs and sperm). The theory of blending inheritance also explained the inheritance of acquired characters (a notion that Darwin did not completely discard). If an organ had been modified during its owner's life, it would produce a modified substance for transmission to the offspring and hence the young would inherit characters not of their parent's youth but of their maturity. It was assumed that the two parental contributions would influence each other, and so the progeny which resulted from this mixture of substances would tend to be intermediate in all characters. Thus a red flower, crossed with a yellow flower, would sometimes produce orange offspring. This would lead to increasing uniformity within the species and not to the range of different characters required for natural selection to work upon. It was difficult to see how favourable variations could accumulate if their effects were diluted constantly in this way. In an attempt to answer this problem Darwin proposed the hypothesis that environmental change brought about a spontaneous production of new variation.

Ironically, the solution to the apparent problem of replenishing variation was discovered in Darwin's lifetime although he was never aware of it. 'Hidden' variation does persist from generation to generation without the need to postulate that it is being generated constantly by outside events. The way in which this occurs will be explained in chapter 2.

'The Descent of Man' and sexual selection

In 1871, Darwin published *The Descent of Man and Selection in Relation to Sex*, really two books in one. Through comparisons between the anatomy, physiology and psychology of man and the great apes, Darwin made the case for the human descent from an ape-like ancestor. Again, bitter controversy ensued. Most of the book is devoted to discussion of what Darwin called *sexual selection*. This operates through some members of an animal species having an advantage, not in personal survival, but in the ability to win mates. Male mammals with effectively threatening weapons, such as the antlers of deer, may succeed in fathering offspring (which will tend to inherit their characters) by driving away male opposition. Attraction of the opposite sex is equally important for successful reproduction; the elaborate plumage and courtship displays of many birds are thought to have evolved because of their importance in a bird's choice of the most outstanding partner to mate with. Characteristics selected sexually need not necessarily be of advantage to their possessors at any other phase of their life.

EVIDENCE FOR EVOLUTION

Darwin marshalled a vast body of facts to support his theory and later investigations have further substantiated his findings.

The fossil record

Fossils, the long-preserved traces of organisms, are powerful direct evidence for evolution because they show similarities, with each other and with living organisms, which indicate a common ancestry. Moreover, for any particular type of fossil, the greater the difference in age between it and its nearest living relative, the more profound are the differences between them. This accords well with the evolutionary idea that organisms diverge from one another and from their ancestors by the gradual accumulation of numerous minute changes.

Fossils form in many ways: occasionally a whole carcase is preserved by freezing, drying or being buried in an acid peat-bog or a tar-pit. Such treatment prevents the agents of decay from decomposing the body. More often, only the hard parts, such as the skeleton or shell of an animal or the woody tissue of a plant, are preserved. These parts last long enough after the decay of the soft parts to become infiltrated by mineral particles and so 'turned to rock'. Perfect fossilisation is rare: often the organism is crushed and distorted and faint impressions are more common than whole fossils. Sometimes tracks of animals are fossilised, as in the case of dinosaur footprints found at Purbeck in Dorset and other parts of the world.

Those organisms which die and become fossilised may become incorporated into layers of sediment which are washed away from eroding rocks and deposited at the bottom of lakes and seas. More recent layers of sediment accumulate on top of older ones so that, after millions of years, a stratified rock is built up with the oldest fossils incorporated into the lower layers and the later ones at the top. Land movements, which may then occur, lift the rock above sea-level where, exposed to air and water, the process of erosion begins again. Millions of years after its death the fossilised organism may be revealed. The fossils which are collected represent only a tiny fraction of those still entombed in the earth, mostly in inaccessible places beneath rocks, deserts, jungles and the sea.

Fossil sequences

Sometimes a whole section of sedimentary rock, from oldest to youngest, is revealed on a cliff edge or a river gorge where the river has cut its way through the layers. This affords the *palaeontologist* (fossil specialist) an opportunity to study strata of known ages and note the changes that occur in the embedded fossilised organisms with the passage of time. Certain sequences of fossils have been carefully recorded and these give abundant evidence for the theory that evolution has occurred.

A well-documented series of changes took place in the structure of a heart-urchin called *Micraster* which is related to the living *Echinocardium*. This urchin is found in several hundreds of feet of chalk beds of the late Cretaceous period (see Figure 2) in England. The oldest and the most recent differ sufficiently for them to be assigned to different species. These extremes are linked by a continuous series of transitional forms which demonstrate gradual changes. For instance, an increase in the number of large spine-bearing knobs on the skeleton suggests (from a study of living urchins) that there was a progressive adaptation for deeper burrowing. The mouth shifts gradually forwards, which may have made feeding more efficient. These are small matters, but they illustrate aptly how the accumulation of minute changes over a long period of time may lead to the creation of a new species. The fact that the strata containing these fossils are on top of one another and the changes are gradual clearly suggests that the representatives of one stratum are descendants of those in the stratum below.

Experience is necessary for the interpretation of such fossil histories, however, since prehistoric earth movements may have caused folding and faulting of the rocks in such a way that layers of a certain age are no longer continuous. Inversions of old layers on top of younger ones may occur.

If fossils come to lie deep in the earth's crust, heat and pressure change the form of the rock containing them and destroy the fossils. So limestone, a richly fossiliferous sedimentary rock, becomes marble which is barren of fossils. Large proportions of the earth's oldest fossils must have been destroyed in this way before man evolved.

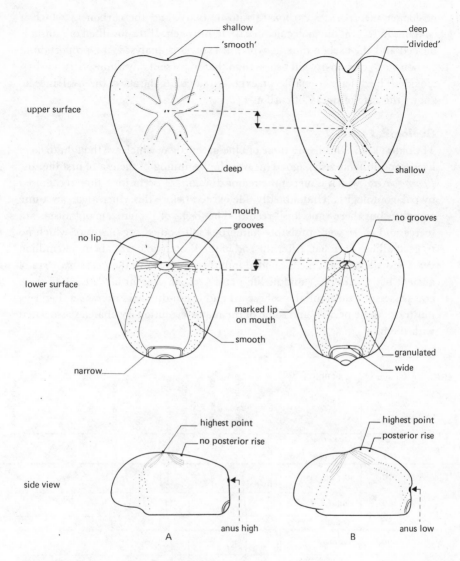

Figure 2 Early and late forms of *Micraster*

Dating rocks and fossils

One method used for dating the layers of sedimentary rocks depends upon the fact that those radio-active elements originally present decay into non-radioactive ones. Thus the proportion of uranium-238 to lead-206, or potassium-40 to argon-40, for instance, indicates the approximate age of the rock. More precise for dating organic remains less than about 50000 years old is carbon-dating. Organic matter is built up by the process of photosynthesis in plants. The plants absorb atmospheric carbon dioxide which contains a fixed, small proportion of the heavy *isotope* (alternative form) of carbon, carbon-14. When the plant (or an animal which has eaten the plant, directly, or indirectly) dies, it no longer incorporates carbon-14 into its body, and the carbon-14 which it contains in its organic molecules decays to nitrogen. If the fossilised organism is burned and the carbon dioxide which it evolves is analysed, the proportion of carbon-14 to carbon-12 will be less than that in present day atmospheric carbon dioxide. The difference enables an expert to calculate the age of the fossil since he knows the rate of decay of carbon-14.

Geological time

The oldest known fossils are those of blue-green algae which are thought to have lived 3200 million years ago. This was the beginning of the era of first life, the *Proterozoic era*, which is one of the major divisions of prehistoric time recognised by palaeontologists. Undoubtedly, life existed before this, since algae are quite complex, but there is no fossil record. The fossils of Proterozoic organisms are scarce and there were probably numerous soft-bodied organisms of which no trace at all has been left. After the end of the Proterozoic era, about 600 million years ago, geological time is divided into three more eras, the *Palaeozoic* ('era of ancient life'), *Mesozoic* ('middle life') and *Cenozoic* ('recent life'). Each of these eras possesses a substantial fossil record and is sub-divided into *periods*. Figure 3 illustrates these periods and some of the major evolutionary changes associated with them.

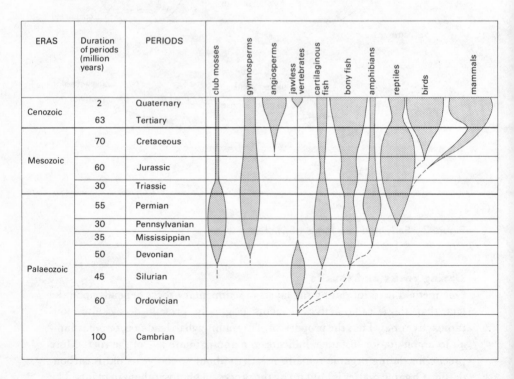

Figure 3 Geological Time Chart showing the relative abundance of some plant and animal groups (after Jenking and Boyce, 1979)

Coal

During the Carboniferous period, 300 million years ago, many events of particular interest and importance to us today occurred. Slow, prolonged earth movements caused some sea-beds to rise whilst land sank. This produced extensive marshes and swamps covering much of Europe and North America. The climate was hot and humid with abundant perennial rainfall. Flowering plants had not yet evolved, but evergreen forests contained thirty-metre high tree-ferns and ten-metre horsetails along with a lush undergrowth of climbing plants. The profuse vegetation died and decayed slowly in the stagnant, marshy ground to become peat. This sank below the surface and sediment accumulated above it. Pressure and heat converted the plant tissue into coal, a substance which retains much of the carbonaceous material of the bodies of the organisms which compose it, so making it useful as fuel. So numerous are fossils in coal that from them much of the higher plant evolution can be traced.

The evolution of the horse

The conditions for fossilisation are exacting and only organisms with suitable hard parts which are covered by sediment under water stand a good chance of being preserved. This means that tracing the evolution of a type of plant or

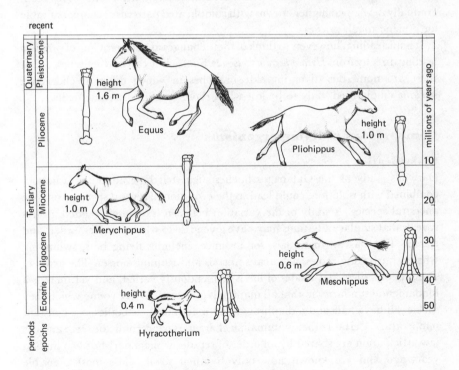

Figure 4   The evolution of the horse
The skeleton of the hind leg is shown in each case. Lateral toes gradually disappear until, in recent times, horses run on the hoof, developed from the nail of one toe. (Not drawn to the same scale)

animal is extremely difficult, except in cases such as that of *Micraster* in which adjacent forms can be compared. Certain evolutionary histories have been compiled from the careful study of fossils from different places, however. The evolution of the modern horse, *Equus*, is such an example. Fossils representing early ancestors of the horse have been found in different continents and, while there can be no direct proof that one form gave rise to another, changes of structure can be arranged in sequence to provide a possible series (Figure 4). Horses are very mobile and it is likely that early kinds would have spread into new continents where there were land connections.

The history of the horse began in North America about 50 million years ago with a cat-sized animal called *Hyracotherium* (previously called *Eohippus*). This animal had low-crowned molar teeth suitable for browsing on soft vegetation. The front feet had four toes each and the hind ones three; this accords with the knowledge that it lived on wooded marshy ground where a wide spread would give support. As the country became drier, *Hyracotherium's* descendants had to survive in open countryside. Fossils of these later forms show that the foot depended more and more on the middle digit which lengthened and thickened. Eventually the foot rested only on the toenail which was now expanded to form a hoof: this aided rapid galloping over hard ground. Meanwhile, the animals became larger, with straighter backs and more complex brains. The teeth gradually developed higher crowns with complicated patterns of enamel suitable for grazing tough grasses.

It is misleading, however, to think of these changes as a straight line of progress leading directly from *Hyracotherium* to *Equus*. Such a line can be disentangled, but there are numerous diverging side branches in which different characters became emphasised, only to be lost when those creatures became extinct.

Comparisons of living organisms

Classification

Darwin's study of the Galápagos finches suggested how adaptive radiation, combined with isolation, could lead to the evolution of several species from one ancestral species. A study of the variation between less closely related groups reveals that similar radiations may have given rise to whole families, orders and classes. The class of mammals, for instance, includes flying bats, swimming whales, burrowing moles, arboreal squirrels and running horses. The array of sizes, shapes, diets and modes of life is tremendously varied, and yet there is a fundamental similarity in that all mammals give birth to live young which they suckle, all have hair and a diaphragm and maintain their bodies at a constant temperature. Certain other mammalian characters concerning the teeth and the jaw articulation are shared by an order of reptiles which flourished 250 million years ago and are known now only by their fossils. The most plausible explanation is an evolutionary one which says that these reptiles gave rise to ancestral mammals which gradually radiated into a variety of different habitats. In the course of time each group became specialised for its particular mode of life and so diverged from parallel groups and from its ancestor.

Plants can be divided into four major types of organisation, called *phyla*, which differ radically from one another. Animals are usually divided into thirteen phyla. Phyla can next be sub-divided into progressively smaller groups, called classes, orders, families, genera and species, between which the common distinguishing features are less and less diverse, kinship being correspondingly closer. The smallest sub-division, the *species*, encompasses not only the smallest structural differences but also the capacity of its members to interbreed.

A scheme of classification upon a basis of kinship and ancestry appears logical with an evolutionary interpretation. Classification is said to be *phylogenetic* where a basis of descent can be demonstrated. As research proceeds, more phylogenetic relationships are revealed which sometimes dictate modifications to the existing scheme of classification. For convenience, arbitrary groupings must be made where phylogenetic relationships have not been revealed.

Comparative anatomy and embryology

Studies of gross structure or *anatomy* of plants and animals reveals patterns of resemblance which are easily interpreted on an evolutionary basis. Resemblances between the *embryos* or unborn young are usually much greater than between adults; very young *vertebrate* (backboned) animal embryos, for instance, are almost indistinguishable (Figure 5). Embryology can also demonstrate

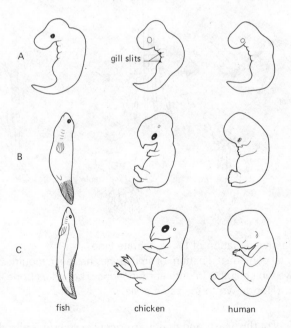

fish   chicken   human

Figure 5   Three comparable stages in the embryology of a fish, a chicken and a human being

The youngest embryos in row A are very similar, while older ones in row B have developed distinctive characters. The oldest embryos, in row C, show recognisable differences peculiar to their species. (Not drawn to the same scale)

unexpected affinities. The stationary jelly-like sea-squirts, which feed on debris filtered from sea-water, used to be classified as relatives of the molluscs until the mobile *larvae* (young forms) were shown to have distinctive features of the group Chordata to which vertebrates belong.

The vertebrates are a particularly well-researched group. Studies convincingly demonstrate that *pentadactyl* (five-fingered) limbs of the amphibians, reptiles, birds and mammals have a fundamental structure in common, a structure they share with the fossil lobe-fin fishes, thought to be the ancestors of the earliest amphibians (Figure 6). Such comparable parts are said to be *homologous*, and are thought to have arisen by adaptive radiation from the ancestral type. Homologous organs provide evidence of affinity between creatures which have undergone 'descent with modification', that is, evolution from a common ancestor.

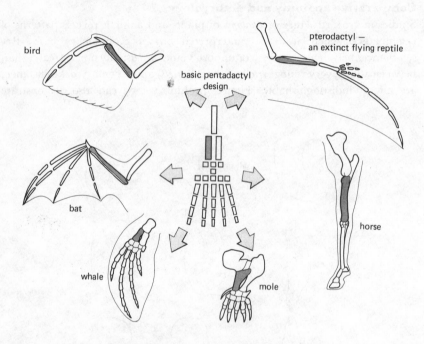

Figure 6   Adaptive radiation of the vertebrate limb
The fundamental pentadactyl pattern of limb-bones has been modified in different groups in ways which suit it to different functions. The radius bone is shaded in each diagram. (Not drawn to the same scale)

The anatomy of the hearts and major arteries of vertebrates also show a basic pattern. The degree of divergence of form between the embryonic single circulation (in which the heart pumps the blood only once during each circuit of the body) and the adult circulation increases in the order fish, amphibians, reptiles, birds and mammals. This reflects the order in time in which these vertebrates are supposed, on the basis of other evidence, to have evolved.

Plants, too, show homologies. Most land-plants possess leaves, appendages of their stems, containing pigments which carry out the food-manufacturing process called *photosynthesis*. Flowering plants possess structurally similar flowers, broadly homologous, in that they nearly all consist of four whorls of elements, carpels (containing 'eggs'), stamens (containing pollen-grains) petals and sepals. The magnolias, which have persisted relatively unchanged for many thousands of years, have leaf-like structures which bear pollen-grains, suggesting that flowers evolved from leaves. Different groups of organisms do not necessarily evolve at the same rate, and organisms, like the magnolias, which continue to bear a 'primitive' character that others have lost, can be a useful aid in interpreting the way in which structures evolved.

Comparative physiology

The complex series of reactions called *respiration* by which living creatures extract energy from their food, is basically the same in all species. Likewise, photosynthetic reactions show great similarities in widely different groups of plants. Most fundamental of all, the substance which encodes all the information for the 'construction' of an organism and is handed down from parent to offspring, functions in all living creatures so far investigated. This substance is called *nucleic acid* and it will be discussed in chapters 4 and 5. Such basic resemblances point strongly to a common origin for all living things. Other physiological similarities, such as the sequence of digestive juices operating in the gut, or the type of blood-pigment, link smaller, more closely related groups.

Interpretations of the evidence of comparative anatomy and physiology must always be made cautiously, however. It is not uncommon for organisms to develop chemical or structural similarities even when their evolutionary origins are quite different. Sometimes the deception is profound: eyes of vertebrates and squids look very similar indeed. A careful study reveals that in the squid eyes the light-sensitive retinal cells face the direction of the incoming light, while in vertebrates the retina is inverted; the sensitive cells lying behind the nerve fibres which carry the message to the brain. Such a seemingly small difference is very significant because it reveals that the two kinds of eye have developed in quite different ways: they are said to be the result of *convergent evolution*. Because they have arisen differently they are said to be *analogous* (similar) rather than homologous.

Distribution of animals

The study of the worldwide distribution of animals supports the evolutionary theory and it is especially revealing where areas once contiguous have been separated by land movements.

Continents

In the northern hemisphere, Eurasia and North America are separated only by the narrow Bering Strait (which geologists have shown relatively recently replaced a land bridge). Eurasian and North American animals show strong resemblances to one another. In the south, the three large land masses of Africa, South America and Australia were all joined to the northern continents at some time in their history and later separated by the sea. Australia has retained its isolation to present times. These southern continents have highly individual faunas; for instance, South America has sixteen endemic families while North America has only four. Australia has the most unusual fauna of all. There are relatively few mammal families and most of these are marsupials or pouch mammals. Here also are the only living *monotremes* or egg-laying mammals, the duck-billed platypus and the spiny anteater.

How do these facts fit in with evolutionary theory? Special creationists argued that the absence of certain types of animals from different continents indicated that they had not been created there because conditions were unsuitable for them. This argument is fallacious, however, since when man has introduced species into new countries they have frequently multiplied so successfully as to threaten the livelihood of the endemic fauna. The more likely explanation is that animals evolved in a certain locality and spread into those habitats which they could reach. The more the descendants were separated from one another in space and time, the greater the divergence between them became. So, those mammals which evolved in the northern hemisphere could have migrated to Africa, South America and Australia since all have had land connections. As the land bridges disappeared, isolating the stocks, the animals evolved in different directions.

Evolution sometimes takes a parallel course in long-separated continents. Thus Australia has marsupial 'moles', 'mice', 'anteaters' and 'flying squirrels' occupying niches similar to those of their placental counterparts in other continents. They are not closely related but come to look alike as a result of their similar ways of life. This is a striking case of convergent evolution.

Islands

Whereas continents have had contact with other land masses in their history, volcanic oceanic islands have always been isolated. A study of their faunal distribution throws particular light on evolutionary theory.

The Azores, the Galápagos Islands and Tristan Da Cunha are examples of oceanic islands. The volcano that threw them from the sea-bed eventually cooled, leaving them lifeless. Gradually they became colonised by species drifting from neighbouring continents in the wind and ocean currents. The colonisers were not wholly representative of the continental fauna and flora from which they derived. This is partly because colonisation is relatively unlikely, particularly if long distances are involved, and partly because certain species are better travellers than others. Turtles manage very well, floating along inside their buoyant shells over vast distances. Birds and insects often fly or are blown. But Darwin noted that oceanic islands are characterised by having few mammals, particularly carnivores, and almost no amphibians. Frogs, when introduced into the Azores by man, multiplied so rapidly as to become a nuisance. They had been unable to get there by themselves simply because amphibian eggs and adults cannot tolerate immersion in sea water. It is nearly always possible to see the evolutionary links between an island's inhabitants and those of the nearest continent but, since the sea provides an effective barrier for most of the time, the island organisms evolve independently. Barriers, including the sea, mountain ranges, deserts or simply an unfavourable temperature range, are essential to the diversification of life.

*

The evidence for evolution increases annually. Darwin assumed that evolutionary change progressed so slowly that evolution by natural (as opposed to human or 'artificial') selection would not be demonstrable within one human lifetime. Now, however, observations on small-scale evolution in the wild, pioneered by such men as E. B. Ford, have led to its becoming a new science. This will be discussed in chapter 7.

QUESTIONS

1 Why do you think that Darwin's theory of evolution was much more widely accepted than that of his predecessors?
2 Take several lines of evidence for and against evolution and use them in an account designed to debate whether the theory of evolution is true. What criticisms of these lines of evidence can you anticipate?
3 Why are islands of such value to an understanding of how evolution occurred?

2 Mendelian Inheritance

THE WORK OF MENDEL

The most famous scientific study of heredity, or genetics, began around 1856 in Brünn, Austria (now Brno, Czechoslovakia) with the work of an Augustinian monk called Gregor Mendel. Mendel chose garden pea plants for his experiments because they occurred in many varieties which 'bred true', that is, by self-pollination they continued to give offspring almost identical to the parents for many generations. Such varieties are called *pure-breeding lines* or *pure lines*. Peas are easy and rapid to cultivate and varieties can be artificially cross-pollinated without difficulty to give fertile offspring.

Mendel's careful studies of peas led him to detect twenty-two contrasted features which possibly could be traced in breeding experiments. To give three examples: peas can have either round or wrinkled seeds, red or white flowers, tall or short stems. Mendel had the insight to appreciate that the mysteries of heredity could be unravelled by studying the inheritance of one pair of characteristics at a time. Previous workers had been confused by results obtained from following the inheritance of several characters simultaneously. Of the characters available for study in peas, Mendel chose seven. For reasons which will be apparent later, it was either an uncannily lucky choice or a remarkably astute one based on a mass of previous experimentation of which we know nothing.

Monohybrid inheritance

Mendel's first recorded experiments were designed to investigate the inheritance of only one pair of characters: what is now called *monohybrid inheritance*. He took pure-breeding tall plants and prevented them from carrying out their normal self-pollination by removing the unripe anthers from the flowers. The tall plant's stigmas were then carefully dusted with pollen collected from short plants and the flowers were enclosed in small bags to prevent other unwanted pollen from reaching them. These tall and short plants are said to belong to the *parental generation*. The seeds ripened and were harvested and planted: they were *hybrids* with a tall mother and a short father. Mendel discovered that they grew, without exception, into plants at least as tall as their maternal parent. They are said to belong to the *first filial*, or F_1 *generation*. Mendel tried the same experiments using tall plants to provide the pollen for the stigmas of short plants (that is, he performed the *reciprocal cross*) and again all members of the F_1 generation were tall. Mendel called the tall character *dominant* and the short character *recessive*.

The tall F_1 plants were allowed to self-pollinate (or *self*) and their seeds were then sown to produce the *second filial* or F_2 *generation*. The F_2 plants were not all the same. Unlike most of his predecessors, Mendel *counted* the individuals with different characters. In one particular cross he recorded 787 tall plants and 277 short ones, that is, approximately three-quarters tall and one quarter short.

Mendel allowed all the F_2 plants to self-pollinate and collected each plant's seeds separately. All the seed collected from short plants bred true, that is, only short plants grew from it. The seed taken from one third of the tall plants also bred true, giving only tall offspring. The seed from the remaining two thirds of the tall plants did not breed true, however, but produced tall and short offspring in the ratio 3:1. The F_2 ratio of 3:1 was in reality a ratio of one true-breeding tall:two non-true-breeding tall:one true-breeding short (see Figure 7). Similar crosses were carried out with the other seven pairs of characters, always with the same result: one character only showed itself in the F_1 generation while the other re-appeared in one quarter of the F_2 generation. At no time did intermediates appear.

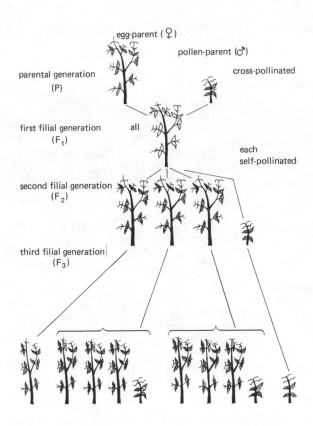

Figure 7 Monohybrid inheritance of height in peas

Particulate inheritance

In Mendel's experiments, the kind of blending of characters that perturbed Darwin did not occur. The shortness of the original short parental pea plants disappeared during the F_1 generation but re-appeared in the F_2. Whatever determined the shortness of these plants was hidden, but not destroyed, during the F_1 generation. Mendel postulated that particles or factors (now called *genes*) which determine characters existed in each parent in pairs and must be transmitted from parent to offspring in the *gametes* (eggs and pollen grains). Each gamete, he reasoned, carries only one of the two parental factors so that at fertilisation the double number is restored in the offspring. The logical conclusion from Mendel's reasoning is that the genes persist unchanged. *Allelomorphs or alleles* are now defined as these alternative genes which have different effects on the same character (Figure 8).

The genetic make-up, or *genotype*, of the pure-breeding tall parent is written TT (each T represents one gene) and that of the short parent tt. When an organism contains a pair of identical genes like this, it is said to be *homozygous*. Each gamete produced by the tall parent can carry only one T gene and those produced by the short parent will carry one t gene each. During fertilisation, T

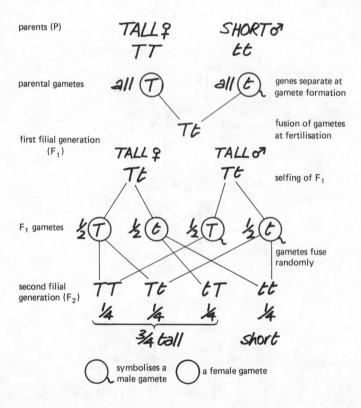

Figure 8 Shorthand representation of a cross between a pure-breeding tall plant and a pure-breeding short plant, carried through to the F_2 generation

and t gametes unite to give offspring with the genotype Tt. Such organisms with two different genes for one character are said to be _heterozygous_.

As Mendel noted, these F_1 heterozygotes look identical to the parental homozygotes because the effects of the 'tall' gene dominate those of the 'short' gene. The parental and F_1 genotypes are different, but their _phenotypes_, or outward appearances, are the same. (The word 'phenotype' covers all detectable aspects of an organism including physiological characteristics such as tolerance to desiccation and resistance to disease.) It is assumed that heterozygotes produce both T gametes and t gametes in equal proportions. When F_1 heterozygotes are selfed, gametes unite at random, regardless of which gene they carry, so that fertilisations TT, Tt, tT and tt occur with approximately equal frequency. A total of three tall to every one short plant is obtained, as Mendel found experimentally.

Monohybrid test-cross

As a check on the equality of numbers of each of the two types of F_1 gamete, Mendel crossed some of the F_1 with the original recessive parent. Tall and short offspring were produced in equal proportions (Figure 9).

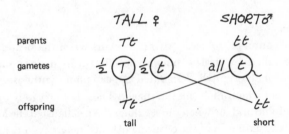

Figure 9   Test-cross

(This is known as a _test-cross_ because it can also be used in situations when we want to test whether an individual showing a dominant character is hetero-zygous or homozygous. Using the symbols Tt and TT, try out the possibilities for yourself.)

Mendel's findings have been summarised by later workers as his _Law of Segregation_ (sometimes called Mendel's First Law). In modern language, this states that characters are determined by pairs of genes which segregate, or pass separately, into gametes. In a heterozygote, the two genes do not alter each other and they are passed on to the offspring unchanged.

Incomplete dominance

A cross between a round radish and a long one produces oval radishes, and a red snapdragon flower pollinated by a white one gives seeds which grow into pink flowers. If the F_1 generations are allowed to self, they will give F_2 ratios of 1 round: 2 oval: 1 long, in the case of the radishes, and 1 red: 2 pink: 1 white snapdragon flowers (Figure 10). Clearly, the heterozygotes are phenotypically

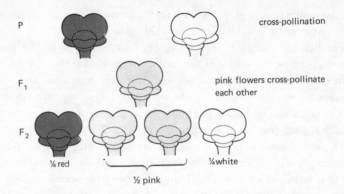

P

cross-pollination

F₁

pink flowers cross-pollinate
each other

F₂

¼ red   ½ pink   ¼ white

Figure 10 A cross between pure-breeding red and white snapdragons followed to the F₂ generation

different from the dominant homozygotes – a situation known as *incomplete dominance*. The genes segregate in exactly the same way as they do when dominance is complete: it is only the phenotype, and not the genotype, which appears to blend.

Lethal genes

Another modification of the basic 3:1 ratio occurs when the bearers of certain genotypes are killed by drastic phenotype effects. In maize, plants with a dominant gene for chlorophyll, designated C–, can photosynthesise. (–indicates that it is irrelevant which allele fills this space.) Those homozygous for a recessive allele, cc, are albino and die when the seedlings have exhausted the food reserves in the seed. About a quarter of the offspring of selfed heterozygotes are therefore without chlorophyll, and what is initially a 3:1 ratio becomes a 3:0 ratio within about a fortnight (Figure 11).

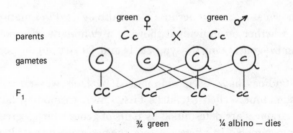

Figure 11 A cross between maize plants heterozygous for a lethal gene

One of the earliest lethals recorded is that of the yellow factor of mice. Black mice, when interbred, produce only black offspring, while black mice crossed with yellow ones produce black and yellow offspring in a 1:1 ratio. This immediately suggests a test-cross between a double recessive and a heterozygote. Yellow × yellow gives two yellow offspring for every one black, not the 3:1 ratio expected if both parents were heterozygous. The explanation is that two yellow genes together kill the embryo early in its development in the mother's uterus. Litters resulting from such crosses are, on average, one quarter smaller than normal. It is interesting to note that, whilst the same gene affects the mouse's coat colour and produces the lethal effects, it is recessive for lethality but dominant for yellowness.

Dominant lethal genes are also known, but in this case they must produce their effects after reproduction has occurred if the gene is to be transmitted to later generations. Huntingdon's chorea, a rare progressive degenerative disease of the human nervous system, is an example. The effects do not normally begin to show until the age of forty or so, by which time the unfortunate sufferers may have already unwittingly passed on the defective gene to their children.

Dihybrid inheritance

Mendel's second task was to ascertain the pattern of inheritance when two characters were considered simultaneously. One of his crosses was between true-breeding plants giving round, yellow seeds with those giving wrinkled, green seeds. Such a cross, in which two pairs of alleles are considered, is called a *dihybrid cross*. It is carried out in the same way as the monohybrid cross.

Mendel's F_1 generation each developed round, yellow seeds, indicating that the round character is dominant to wrinkled and yellow is dominant to green. After the fifteen F_1 plants used had selfed, Mendel counted 556 seeds of the F_2 generation – of these there were:

315 round and yellow
101 wrinkled and yellow
108 round and green
32 wrinkled and green.

These different kinds of F_2 peas were frequently present in one pod.

It was obvious that the roundness of the original parent seeds was in no way tied to their yellowness, since both round green and wrinkled yellow F_2 seeds appeared. These new combinations are called *recombinant types* to distinguish them from the original *parental types*. Mendel had demonstrated that the seed colour factor segregated independently of the one for seed shape.

Taking each character separately, it is clear that approximately $\frac{3}{4}$ of the F_2 seeds were round and $\frac{1}{4}$ wrinkled; $\frac{3}{4}$ were yellow and $\frac{1}{4}$ green. Considering the characters together, they approximate to a ratio of 9 round yellow: 3 wrinkled yellow: 3 round green: 1 wrinkled green. Mendel found that this 9 : 3 : 3 : 1 ratio occurred also both as a result of the reciprocal cross and when the original parents were round and green crossed with wrinkled and yellow, that is, when each parent carried one dominant character. Other combinations of different phenotypes which Mendel investigated always gave the same F_2 ratios.

In genetic symbols, the true-breeding parent plants must be RRYY and rryy, that is, homozygous for both pairs of alleles. Each gamete carries only one gene of each pair and so they can be represented as RY and ry respectively. All the F_1

offspring are therefore heterozygous for both pairs of alleles, RrYy. Since all possible combinations of offspring appear in the F_2 generation, it follows that four different sorts of gametes are produced by the F_1, thus: RY, Ry, rY and ry, and that these combine at random. To show all sixteen combinations by means of the kind of diagram in Figure 8 would be confusing, and so a check-board can be used (see Figure 12).

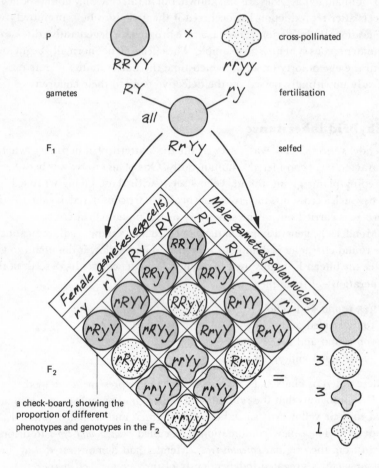

Figure 12 A diagrammatic representation of the independent assortment of two pairs of alleles in peas

Dihybrid test-cross

The plants grown from the round, yellow F_1 seeds can be used as parents in a test-cross with *double recessive* plants (that is, those which are recessive in both characters under consideration). The four different phenotypes are produced in equal proportions, which is what the Mendelian theory predicts.

These results are summarised by Mendel's second law, the *Law of Independent Assortment* which says that either of a pair of alleles can combine with either of another pair.

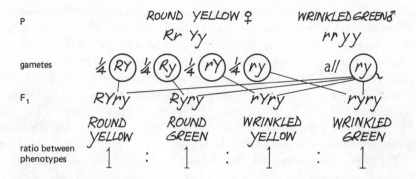

Figure 13 Dihybrid test-cross

Results of Mendel's Work

Mendel clearly knew what to expect before he performed his elegant experiments. In choosing single characters to study at a time and keeping careful numerical records of crosses, he was able to draw valuable general conclusions which have not been seriously challenged since.

Mendel wrote a paper summarising his work, entitled 'Experiments in Plant Hybridisation' which he read in 1865 to the Brünn Natural Science Society. It was published the following year in the 'Proceedings' of the Society. Reprints of Mendel's paper were received by a number of scientific libraries in Europe, including that of the Linnaean Society in London, but, oddly, no one working in the field comprehended it sufficiently to make use of it. Mendel tried to interest Karl Nägeli, a distinguished Swiss botanist, in his ideas but Nägeli, too, failed to grasp their significance. Darwin (whose *Origin of Species* Mendel had found stimulating), with sixteen years still to live, remained in ignorance of this vital solution to a problem which puzzled him.

It was not until 1900, after Mendel's death, that his principles were rediscovered by three scientists who had been working independently on breeding experiments similar to Mendel's: von Tschermak in Austria, Correns in Germany and de Vries in Holland. It is not entirely clear whether they read Mendel's paper before or after their own discoveries, but all three acknowledged Mendel's work as having precedence, and his paper took its place amongst the most important biological literature.

During the following ten years, Mendel's work was successfully repeated and his laws found to apply to other species of plants and to animal species.

PROBABILITY

The F_2 phenotypic ratios of 3 : 1, obtained in a monohybrid cross, and 9 : 3 : 3 : 1, resulting from a dihybrid cross, are seldom exact. The first is a way of saying that for every plant which develops there are three chances in four that it will be, say, tall and one chance in four that it will be short. The chance element occurs in two different ways. Firstly, when the genes separate (or segregate) into the gametes, it is possible that a few more carrying the 'tall' gene (that is, the gene which determines the tall character) than carrying the 'short' gene (or vice versa) will develop to maturity. Of those which mature, it is only a minority which actually effect fertilisation, and they do so quite irrespectively of which allele they contain. Secondly, fertilisation being purely random, it is possible that more t egg cells are fertilised by, say, T pollen grains than by t pollen grains to form the F_2 generation. This chance deviation from expected ratios will be more pronounced in small samples than large ones.

The same principle applies when tossing a coin. The theoretical chance of getting heads is fifty per cent, or 0.5 : 1 but, if the coin is tossed only twice, we would not be unduly surprised to find it landing heads on each occasion. It would be far more unexpected to obtain 100 heads from 100 tosses; since the sample is now much larger, the actual result almost always matches the expected result more closely. The difference between observed and expected figures is called the *sampling error*.

Rules of probability

Addition
If two events are mutually exclusive (for example, a pea plant may be tall or short but not both) and p_1 is the probability of one occurring and p_2 is the probability of the other, then the probability (p) that one or the other will occur is $(p_1 + p_2)$. That is, it is certain that the pea plant will be either tall or short. The general rule for any number of mutually exclusive events is:

$$p = p_1 + p_2 + p_3 + \ldots = 1$$

where 1 stands for 100 per cent, or all possible cases.

Multiplication
If two events are independent (for example, the colour of Mendel's peas did not affect their shape) then, if p_1 is the probability that the first will occur and p_2 the probability that the second will occur, $(p_1 \times p_2)$ is the probability that they will both occur together. If the probability of yellow seeds in the F_2 generation is 0.75 and the probability of round seeds is also 0.75, the probability of round, yellow seeds is:

$$0.75 \times 0.75 = 0.5625$$

or
$$\tfrac{3}{4} \times \tfrac{3}{4} = \tfrac{9}{16}$$

In general:
$$p = p_1 p_2 p_3 \ldots = 1$$

Now, in estimating whether or not figures obtained experimentally are in close accordance with theoretical predictions, we must make allowance for sampling error. It is conceivable that we might toss a coin 100 times and obtain 100 heads but it is so unlikely that if it were to happen we would suspect that there was a bias in the coin. The error would be too great for us to accept that our hypothesis (that the coin should fall heads in fifty per cent of the tosses) was correct. Similarly, in breeding experiments, too wide a divergence from an expected 3 : 1 ratio would lead us to suppose that the experiment suffered from errors or else our prediction about the expected ratios was incorrect. All that remains is to define how much departure from the expected results we are prepared to accept as mere sampling error.

If the observed results diverge from the expected ones by an amount no greater than what we could anticipate once in twenty occasions (or more often) by chance alone, then we normally assume that the divergence *is* pure chance. We say that the divergence is not significant and we retain our original hypothesis. If a divergence is so large that it would occur, say, only once in a hundred tries by chance, then we can reasonably expect that such a wide divergence will *not* happen by chance. If we found such a divergence experimentally, we would suspect that the hypothesis was wrong. It is important to see, however, that this is calculated guesswork, as once in a hundred times a divergence of this size *would* occur by sheer chance without the hypothesis being faulty. If the consequences of wrongly rejecting a hypothesis would be serious (in making an important medical decision, for instance), then we might choose to be cautious and accept divergences of this magnitude. If we did so, we would also increase our chances of accepting a false hypothesis.

The Chi-squared test

The Chi-squared (χ^2) test is a method of analysing results to estimate the degree of departure from expected ratios (χ is a Greek letter, not an English X, and is pronounced 'kai'). It applies accurately only when the number of trials is large, ideally a hundred or more.

χ^2 is defined as:
$$\sum \frac{(O-E)^2}{E}$$

where O is the observed frequency of an event and E is the theoretical or expected frequency. $(O-E)$ is the divergence between one observed result and the corresponding expected one. Squaring it ensures that the value is positive. Σ is a summation sign which indicates that the different classes of results should be added together.

An example from Mendel's results

In a monohybrid cross, Mendel discovered that when plants pure-breeding for yellow unripe pods were crossed with those pure-breeding for green unripe pods, all the pods of the F_1 generation were green. Allowed to self, they gave rise to an F_2 in which 428 pods were green and 152 yellow. This gives a ratio of 2.82 : 1 when, of course, the predicted ratio is 3.00 : 1.

The expected numbers in each class are found by simple arithmetic, thus:

Total number of pods $= (428 + 152) = 580$

Expected number of yellow pods $= \dfrac{580}{4} = 145$

Expected number of green pods $= (145 \times 3) = 435$

Using the Chi-squared symbols

Table 1

Colour of pod	O	E	$O-E$	$(O-E)^2$	$\dfrac{(O-E)^2}{E}$
Green	428	435	-7	49	$\dfrac{49}{435}$
Yellow	152	145	$+7$	49	$\dfrac{49}{145}$

$$\chi^2 = \frac{49}{435} + \frac{49}{145} = 0.113 + 0.338 = \underline{0.451}$$

All that remains is to find out how often a χ^2 of 0.451 will occur by chance alone. The answer is obtained by consulting a table of Chi-squared.

Table 2   Table of χ^2

Degrees of freedom, N	Probability, P										
	0.99	0.98	0.95	0.90	0.75	0.50	0.25	0.10	0.05	0.02	0.01
1	0.00	0.00	0.00	0.02	0.10	0.45	1.32	2.71	3.84	5.41	6.64
2	0.02	0.04	0.10	0.21	0.58	1.39	2.77	4.61	5.99	7.82	9.21
3	0.12	0.19	0.35	0.58	1.21	2.37	4.11	6.25	7.82	9.84	11.34
4	0.30	0.43	0.71	1.06	1.92	3.36	5.39	7.78	9.49	11.67	13.28
5	0.55	0.75	1.15	1.61	2.67	4.35	6.63	9.24	11.07	13.39	15.09

It is necessary to know the number of 'degrees of freedom' operating. This is usually one less than the number of classes involved, and represents the number of *independent* classes. In this case there are two classes, yellow pod and green pod, and fixing the number of one of them from a given total automatically determines the other.

In Table 2, then, with one degree of freedom, the value of χ^2 (or else two which bracket it) is found. The figure 0.45 appears in this column, and this is very close to 0.451. Reading up to values of probability, P, $\chi^2 = 0.45$ corresponds to a probability value of 0.50. This means that a deviation from the 3:1 ratio as large as Mendel obtained for this cross could be expected in approximately fifty per cent of the trials. In other words, the deviation is not significant and is very likely to be due to chance. There is no hesitation in accepting the hypothesis on which the experiment was based. We say there is a *good fit* between the observed results and the expected ones.

All Mendel's observations match the expected frequencies so closely that there has been a suggestion that he (or his assistants) might have 'improved' the results a little in order to give his hypotheses greater credence!

The 5 per cent (or one in twenty) probability level is the usual limit of acceptance, so that a χ^2 is accepted as being within the normal chance range of variation when it falls between $P = 0.99$ and $P = 0.05$ on the Chi-squared table.

PROBLEMS OF EXPLAINING CONTINUOUS VARIATION

The kinds of patterns of inheritance which commonly occur appear to bear little relationship to Mendelian ratios. This does not mean that they are obeying different laws: on the contrary, the principles of Mendelism have been confirmed as virtually universal. But the result of many natural crosses is a series of offspring which vary *continuously*, that is, the dimension of a character may have any value within a certain range. (The sharply defined characters studied by Mendel were *discontinuous*.) Sons' and daughters' heights rarely bear any exact relationship to those of their parents, for instance.

Environmental effects

One of the reasons for continuous variation is that all phenotypic characters are influenced, to varying degrees, by the effects of the environment. A genotypically tall child may be dwarfed by starvation and so appear similar to one whose genotype is for shortness. No character of any organism can be said to be solely due to the effects of the environment or solely due to heredity: both are always operating. In the case of continuously varying characters, such as human height, the genotype determines the phenotypic *range* within which an individual will fall and the environment determines the exact point in the range. In controlled breeding experiments it is essential to keep the environmental conditions as constant as possible so that observed variations can be attributed to genotypic differences.

Polygenic inheritance

Shortly after the rediscovery of Mendel's work, it was suggested that continuous variation could be explained if several genes affected one aspect of the phenotype. For example, in a hypothetical case, supposing two dominant genes, A and B, both determine dark coat colour in a mammal and their effects are additive. Their recessive alleles, a and b, produce white coats. The darkness of any individual's coat will depend upon the number of A and B genes in the genotype. Figure 14 illustrates a cross between individuals AABB and aabb followed through to the F_2 generation. In the F_2, $\frac{1}{16}$ of the offspring have four dominant alleles, $\frac{4}{16}$ have three, $\frac{6}{16}$ have two, $\frac{4}{16}$ have one and $\frac{1}{16}$ have none. If the numbers of dominant alleles, and hence the darkness of the coats, are expressed as a histogram, a distribution is obtained which resembles a normal curve.

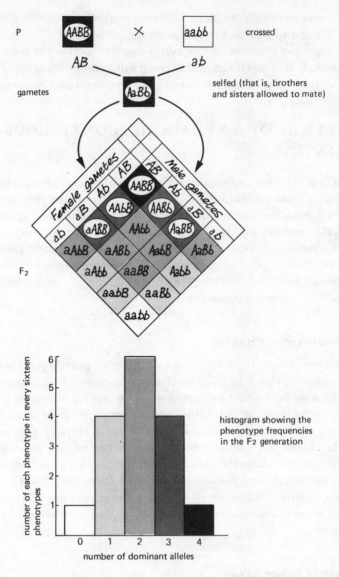

Figure 14   A simple, hypothetical case of polygenic inheritance of coat colour in mammals

If a larger number of genes were involved, there would be finer grades of darkness in the F_2 generation and the parental types would re-appear more infrequently. Work out how many different F_2 phenotypes would result if the dark character were determined by four, and then by six, pairs of alleles. Such *polygenic inheritance* (that is, in which the character is determined by several genes) has been demonstrated in many cases, including the inheritance of length of cob of maize plants and that of human skin colour. The issue of polygenic inheritance is further complicated by the fact that different dominant genes may affect the phenotype to different degrees.

Analysing the effects of genes which have small additive effects is a tedious business requiring careful statistical analysis. In all cases where this has been done, however, the conclusion has been that, while the phenotypes appear to blend, the genes behave according to the rules of Mendelism.

PEDIGREES

The study of human inheritance is complicated by the impossibility of planning particular breeding experiments. Geneticists must make the best of analysing pedigrees for certain traits. Despite the difficulties, some characters associated with single pairs of alleles have been fully documented. Albinism, the condition in which the body does not develop melanin pigment, giving white hair, pale skin and pink eyes, is one example. The gene for albinism is recessive, so that if an albino child is born to phenotypically normal parents it reveals that they are both heterozygous carriers of the gene.

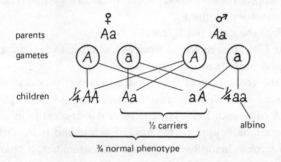

Figure 15   Inheritance of albinism

There is a one-in-four chance that any subsequent child the couple had would be similarly affected.

A minor difference in human ear-lobes follows a similar inheritance pattern. A pedigree for 'free' and 'attached' lobes is shown in Figure 16. Work out what the different genotypes must be.

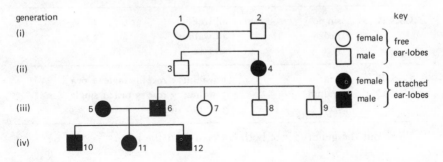

Figure 16   Pedigree of human ear-lobe shape (the numbers are for identification of individuals)

Some disorders are associated with the transmission of single dominant genes. Polydactyly (extra fingers and toes) and one form of dwarfism are examples.

QUESTIONS

1 In certain flowers, the cross, red × yellow, gives red and yellow offspring in a 1:1 ratio, but, red × red, gives rise to red only.
   (a) Which colour is dominant?
   (b) What are the red and yellow plants' genotypes?

2 Two curly-winged fruit-flies are mated and the F_1 consists of 341 curly-winged flies and 162 normal ones. Explain.

3 Pure-breeding peas with grey seed-coats were crossed with those with white coats. The cross was followed in the usual way to the F_2 generation in which 704 seeds were grey and 224 were white. Assume that this difference is controlled by a single pair of genes.
   (a) Using appropriate symbols, make diagrams of the crosses from the parental generation to the F_2.
   (b) Calculate the expected F_2 results.
   (c) Use the Chi-squared test to work out whether the experimental results deviate significantly from the expected ones.

4 In cattle, the cross, horned × hornless, sometimes produces only hornless cattle whereas, in other such crosses, horned and hornless appear in equal numbers. A cattleman has a large herd of hornless cattle in which horned progeny occasionally appear. He has red, white and roan (red-and-white) animals, the roans being heterozygotes for coat-colour, a character which expresses no dominance. The cattleman wishes to establish a pure-breeding line of red, hornless animals. How should he proceed?

5 In the domestic fowl, comb-shape is determined by two pairs of alleles, P and R, which interact with one another to produce the phenotype. Two cocks, A and B, with rose-shaped combs, each mate with three hens. The results of the crosses are shown below:

	Hen's comb-shape	Ratio of offsprings' combs
Cock A	single	all rose
	pea	all walnut
	rose	all rose
Cock B	pea	1 walnut : 1 rose : 1 single : 1 pea
	walnut	3 walnut : 3 rose : 3 pea : 1 single
	rose	3 rose : 1 single

Work out the genotypes of both cocks and all the hens.

3 Chromosomes

By the time the principles of Mendelism had been rediscovered, the science of studying cells, or *cytology*, was well established. The development of the light microscope made great advances possible and details of cell structure were elucidated throughout the nineteenth and early twentieth centuries. More recently, the electron microscope, with its ability to resolve objects under 0.01 micrometre, has been a useful aid to the increase of knowledge in this sphere.

Threads of material which readily take up basic (i.e. non-acidic) stains, the *chromosomes*, had been observed in cells as early as 1848 by Wilhelm Hofmeister. Later workers noted that each cell of a particular species had a constant number of chromosomes and observed the regular way in which these duplicated and divided prior to ordinary cell divisions. The halving of chromosome numbers in gamete-formation was also known. Cytologists appreciated the implication that chromosomes are essential to heredity.

Today we know the chromosome numbers of many species: for instance, man has 46, the fox 34, red clover 14 and maize 20. It is no coincidence that these are all even numbers. Every individual formed as a result of sexual reproduction inherits a set of chromosomes from its mother (called *maternal chromosomes*) and a similar set from its father (*paternal chromosomes*). The human egg, for example, contains 22 ordinary chromosomes or *autosomes* and a *sex-chromosome* designated X. The sperm contains 22 similar autosomes and either an X or a Y sex-chromosome. The total of 23 chromosomes per gamete is called the *haploid number* and is given the symbol n. At fertilisation, the maternal and paternal sets of chromosomes are brought together in the nucleus of the zygote. This cell therefore contains two sets of chromosomes which together give the *diploid number*, $2n$. In humans, $2n = 46$. The zygote, and all the cells that develop from it in the formation of the adult organism, are called diploid cells.

CHROMOSOME STRUCTURE

The chromosomes described here belong to organisms in which a nuclear membrane contains the chromosomes. The genetic material of bacteria, viruses and blue-green algae is not organised into 'true' chromosomes, nor is it enclosed within a nuclear membrane.

Chromosomes of a non-dividing cell are extremely long and thin and dispersed in the fluid nuclear sap. Certain portions of chromosome may be more condensed than the rest, forming granules which readily retain dye and may therefore be rendered visible. Almost all non-dividing nuclei contain one or more oval *nucleoli*, each of which is attached to a specific region of chromosome called the *nucleolar organiser*.

Chromosomes become much shorter and thicker and hence readily visible under a light microscope, only when they are undergoing replication and division. At maximum contraction, each resembles a rod with a conspicuous constriction, the *centromere*, at some characteristic point along its length. The centromere positions and the particular lengths and shapes of chromosomes in a haploid set allow them to be individually identified in some species.

Chemically, chromosomes consist of a *nucleic acid*, called *deoxyribonucleic acid* (DNA), and protein. A DNA molecule comprises two complementary strands linked together in the form of a double helix. Each chromosome is thought to contain just one DNA molecule, much coiled and supercoiled. The nucleus of one human cell contains 7.3 picograms (pg) of DNA distributed amongst its 46 chromosomes. (One picogram is 10^{-12} gram.) Each picogram of DNA, as a single molecule, stretches 31 cm and so a human cell contains about 2.26 metres of DNA. A human chromosome averages 4 to 6 micrometres (1 micrometre is 10^{-6} m) long; clearly some intricate packaging of the DNA must occur.

The chief type of protein associated with chromosomes is *histone*, a basic protein, except in some sperm cells in which *protamines* are found instead. Certain fungi have no histones in their chromosomes; in some ways they are intermediate between the 'higher' organisms, with chromosomes as described here, and bacteria, viruses and blue-green algae.

MITOSIS

An adult human being is constructed of approximately 10^{14} cells, all derived from a one-celled zygote. The zygote undergoes a series of nuclear divisions, each called *mitosis*, in which every chromosome duplicates itself exactly and then the two identical sets of chromosomes separate. The zygote's cytoplasm divides after the first nuclear division so that equal amounts surround the two new nuclei. One diploid cell therefore turns into two diploid cells and the process repeats rapidly as the embryo grows. Details of mitosis may be found in *The cell concept* in this series.

MEIOSIS

If gametes were diploid cells, fertilisation would result in a zygote that was *tetraploid* (*4n*) and the number of chromosomes per cell would double each generation. In 1887, a German zoologist, August Weismann, predicted that a reduction division would be detected; this would precede gamete-formation and halve the numbers of chromosomes in gametes. Such a division was later discovered and found to be universal in sexually-reproducing organisms (with the exception of certain micro-organisms). We now call it *meiosis*.

The essential elements of a meiotic division are the duplication of the diploid number of chromosomes followed by two divisions which separate the chromosomes into four haploid gametes. Meiosis and fertilisation are therefore two complementary requisites for sexual reproduction (Figure 17).

Meiosis is described in a number of stages, for convenience, but the process is continuous (see Figures 18 and 19).

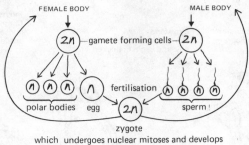

Figure 17   The effects of meiosis and fertilisation on chromosome numbers in animals such as humans
During meiosis, each cell becomes temporarily tetraploid (4n) before undergoing two divisions to form two haploid gametes

Pachytene of prophase

Diplotene of prophase I

Diakinesis of prophase I

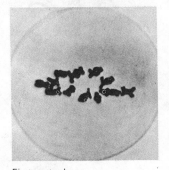

First metaphase

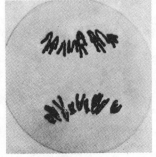

First anaphase

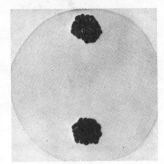

First telophase

Figure 18   The main stages in meiosis

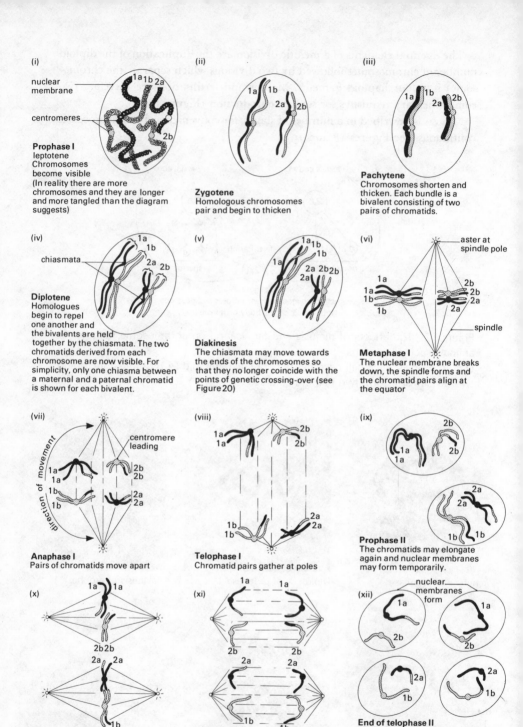

(i)

nuclear membrane

centromeres

1a 1b 2a

2b

Prophase I
leptotene
Chromosomes become visible
(In reality there are more chromosomes and they are longer and more tangled than the diagram suggests)

(ii)

1a 1b

2a 2b

Zygotene
Homologous chromosomes pair and begin to thicken

(iii)

1a 1b

2a 2b

Pachytene
Chromosomes shorten and thicken. Each bundle is a bivalent consisting of two pairs of chromatids.

(iv)

chiasmata

1a
1b

2a 2b

Diplotene
Homologues begin to repel one another and the bivalents are held together by the chiasmata. The two chromatids derived from each chromosome are now visible. For simplicity, only one chiasma between a maternal and a paternal chromatid is shown for each bivalent.

(v)

1a 1b
1b

1a

2a 2b 2b

Diakinesis
The chiasmata may move towards the ends of the chromosomes so that they no longer coincide with the points of genetic crossing-over (see Figure 20)

(vi)

aster at spindle pole

1a
1a
1b

2b
2b
2a

2a
1b

spindle

Metaphase I
The nuclear membrane breaks down, the spindle forms and the chromatid pairs align at the equator

(vii)

direction of movement

centromere leading

1a
1a

2b
2b

1b
1b

2a
2a

Anaphase I
Pairs of chromatids move apart

(viii)

1a
1a

2b
2b

1b
1b

2a
2a

Telophase I
Chromatid pairs gather at poles

(ix)

2b

1a
1a

2b

2a
2a

1b
1b

Prophase II
The chromatids may elongate again and nuclear membranes may form temporarily.

(x)

1a 1a

2b 2b
2a 2a

1b

Metaphase II
Two spindles form at right angles to the earlier one

(xi)

1a 1a

2b 2b

2a 2a

1b 1b

Anaphase II
The centromeres divide and sister chromatids separate

(xii)

nuclear membranes form

1a

1a

2b

2b

2a

2a

1b

1b

End of telophase II
A tetrad of four nuclei, each with the haploid number of daughter chromosomes

Figure 19   Stages in meiosis

First meiotic division

Prophase I

During the course of prophase the nucleolus disperses, the two *centrioles* (short rods with a characteristic arrangement of internal fibres, found in the cells of animals and some plants) migrate to opposite sides of the nucleus and the chromosomes condense. Prophase is longer and more complicated than the equivalent stage in mitosis and substages are recognised:

1 Leptotene

The diploid number of chromosomes becomes visible as long, slender threads which will now absorb certain stains. Characteristic, irregular swellings, called *chromomeres*, are present in each thread. Each thread appears single but it consists, in fact, of a tightly twisted pair of threads called *chromatids*, since duplication of the chromosome material has already occurred. The centromeres remain undivided.

2 Zygotene

Each chromosome locates its homologous partner and the two lie closely together, matching point for point in a process called *synapsis*. There are thus four strands in each group, twisted around each other, although normally only two strands (corresponding to the paternal and maternal chromosomes) can be seen. The forces of attraction between homologous chromosomes are not fully understood.

3 Pachytene

The chromosome bundles (called *bivalents*) shorten and thicken by coiling up.

4 Diplotene

By the end of pachytene, the attraction between homologous chromosomes finishes and they tend to separate. Complete parting is prevented because, at certain places in each bivalent, two chromatids, one derived from the paternal and one from the maternal chromosome, form a cross. Each cross is a *chiasma* and long chromosomes may have several chiasmata, giving the appearance of a series of loops. Other chromosomes develop only one chiasma and form an X-shape. With rare exceptions, every normal bivalent has at least one chiasma; they seem to hold the bivalent together. The interpretation of the chiasma is that two chromatids derived from different chromosomes have previously broken and rejoined to the 'wrong' ends. The breakage and fusion gives rise to the appearance of a cross, since the mutual attraction between portions of *sister-chromatids* (those derived from one chromosome) makes them lie as close as possible together, while portions of non-sister chromatids tend to repel one another. The chiasma is the visible evidence that the exchange of genetic material between two non-sister chromatids has occurred. This failure of linkage, or *crossing-over* has important genetic consequences (see p. 42). For interpretation of the chiasma, see Figure 20.

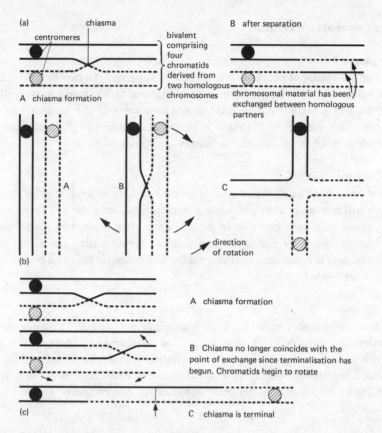

Figure 20(a) Representation of the exchange of chromosomal material which results from chiasma-formation. (b) An explanation of the formation of cross-shaped bivalents after chiasma-formation. Sister-chromatids attract one another, while homologues (differently shaded) repel. The resulting rotation of the chromosomes brings the centromeres as far apart as possible. (c) Diagram illustrating terminalisation

5 Diakinesis

The transition from diplotene to diakinesis involves the gradual shortening and thickening of the chromosomes. In some species, the chiasmata appear to progress towards the ends of the chromatids so that the visible cross no longer corresponds to the point of genetic exchange (Figure 20(a)). This process of *terminalisation* may proceed until the two pairs of chromatids are held together only at their ends.

At the end of prophase, the nuclear membrane disappears and a *spindle*, a fibrous proteinaceous structure appears. Where centrioles are present, these give rise to radiating systems of fibres called *asters*, one at each spindle *pole* (the 'end' of the spindle).

Metaphase I

The bivalents become attached to the spindle in such a way that homologous centromeres are on opposite sides of the widest part of the spindle, the *equator*, and equal distances from it. Chromatids derived from a maternal chromosome may be 'above' or 'below' the equator at random, so that each half of the spindle bears chromatids of paternal and maternal origin.

Anaphase I
The centromere of each half-bivalent begins to move apart from the homologous centromere, aided by elongation of the spindle, and each drags its trailing chromatids towards the nearest pole. There may be some resistance caused by the entanglement of chromatids which have not completed the process of terminalisation.

Telophase I
The random arrangement of half-bivalents at metaphase means that those which accumulate near each pole are new combinations of mixed paternal and maternal origin. In some organisms, the groups of telophase half-bivalents may pass into *interphase*, a stage during which the chromatids become long and invisible. In other organisms, the second prophase begins immediately.

Second meiotic division

Prophase II
The paired chromatids (half-bivalents) become visible again, attached only by their common centromeres. There is a superficial similarity to the prophase of mitosis, but there are significant differences. Firstly, there is only a haploid number of paired chromatids whereas, during the prophase of mitosis, the diploid number is present. (Mitosis occurs in certain haploid cells, however, in which case only the haploid number of paired chromatids is present.) Secondly, during meiosis the chromatids do not coil about one another tightly as they do in mitosis. Thirdly, and importantly, as a result of exchanges during chiasma formation, the chromatids which are joined by a common centromere are no longer identical.

Two spindles appear, usually oriented at right angles to the direction of the first division spindle.

Metaphase II
The centromeres and their attached pairs of chromatids arrange themselves on the equator of the spindle.

Anaphase II
The centromeres divide and migrate to opposite poles, each dragging behind it a chromatid (now a *daughter chromosome*).

Telophase II
Nuclear membranes form around each of the four haploid groups of chromosomes. The chromosomes revert to their long, non-stainable state. The result of this final division is to produce a *tetrad* of four haploid cells.

Cytoplasmic division

In the male animal, cytoplasmic division follows in a manner similar to that accompanying mitosis, giving four equal cells which mature to become sperm. In the female animal the cytoplasmic divisions are unequal, so that three sets of chromosomes are enclosed in a minimum of cytoplasm. These are called *polar bodies* and do not survive. The fourth of the group retains all the rest of the cytoplasm and becomes a functional egg.

In plants, the gamete-forming tissues undergo meiosis in the same way as those of animals. A female *megaspore mother-cell* gives rise to four *megaspores*, three of which degenerate while the remaining one becomes the egg. In the anthers, four functional *microspores* are produced by meiosis of pollen mother-cells.

THE CHROMOSOMAL THEORY OF HEREDITY

According to Mendelian theory, a body cell (or *somatic cell*) contains twice as many genes as a gamete. By 1900, cytologists knew that chromosomes were carefully replicated during cell division and that their number was halved in the formation of gametes. Some research workers came to the conclusion that genes were borne on chromosomes. This was a profound step forwards since, previously, it had been generally accepted that genes could not have a material existence at all. In 1903, Sutton proposed a modern interpretation of the relationship between genes and chromosomes and explained the Mendelian principles of segregation and independent assortment in terms of chromosome behaviour. The amalgam of genetics and cytology gave birth to a fruitful new science called *cytogenetics*.

The segregation of genes and the separation of chromosomes during meiosis lead to the conclusion that homologous chromosomes carry pairs of alleles. Consider an organism with only three pairs of homologous chromosomes, each pair bearing a pair of dissimilar alleles, Aa, Bb and Cc. One allele of each pair is derived from the father and one from the mother. During meiosis, these paired alleles separate into different gametes, obeying Mendel's Law of Segregation. The arrangement of bivalents at the equator is random, however, so that there is no particular preference for the alignment of chromosomes as shown at metaphase I in Figure 21 (a). Figure 21 (b), (c) and (d) show three other equally probable arrangements. The consequences are that eight kinds of gametes carrying different allele combinations are produced in equal proportions. This is the physical basis for Mendel's Law of Independent Assortment: each of a pair of alleles may combine with either of another pair. Most organisms contain more chromosomes than just three pairs. The number of possible different chromosome arrangements in the gametes is given by 2^n, where $n =$ the haploid number.

Although the view that genes are borne on the chromosomes is based on strong evidence, it is circumstantial evidence from two separate lines of investigation.

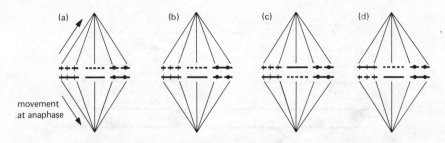

movement
at anaphase

Figure 21　Diagram to show the possible arrangements of chromosomes at the equator of the spindle during metaphase I

Only three pairs of homologous chromosomes are shown, one of each pair dotted and one continuous. Their division into paired chromatids is not represented. Three homologous pairs can be arranged in four ways, as shown, to give eight different complements of chromosomes at the poles after anaphase I. Most organisms contain far more than three pairs of chromosomes per cell, so the number of different groupings resulting from random orientation on the spindle is much greater

LINKAGE

Sutton perceived that it was likely that a chromosome would carry more than just a single gene, and that the genes belonging to a particular chromosome would be inherited together. In 1905, Bateson and Punnett produced some data which strongly supported Sutton's theory.

Bateson and Punnett studied the inheritance of two pairs of alleles in the sweet-pea. Pure-breeding plants with purple flowers (P) and long pollen-grains (L) were crossed with red flowers (p) having round pollen-grains (l). The F_1 generation flowers were all purple with long pollen, since these characters are dominant. In the F_2, however, there was much deviation from the expected 9:3:3:1 ratio, with far more of the parental kinds (purple long and red round) than recombinant kinds (purple round and red long). When the experiment was repeated using purple, round pollen and red, long pollen plants as parents it was *these* kinds which predominated in the F_2.

It is now clear that the genes which determine pollen shape and flower colour are situated on the same chromosome, that is, they belong to the same *linkage group*, but that linkage is incomplete so occasional re-arrangements occur. An organism's number of linkage groups equals its haploid number. This is further powerful evidence to suggest that the genes belonging to one linkage group (a genetical concept) are arranged on one chromosome (a cytological concept). If the genes determining the two dominant characters are on one chromosome, while the genes determining both recessives are on its homologue, the linked genes are said to be *in coupling*. If each of a pair of chromosomes carries one of the dominant genes, these are *in repulsion*.

It is obvious from Bateson and Punnett's results that such linkage is not necessarily permanent, since they noted a low frequency of recombinants. If the

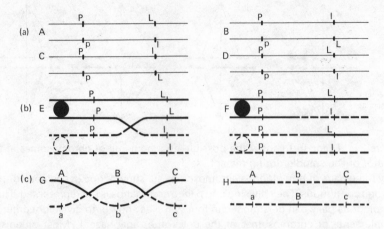

Figure 22(a) Diagram illustrating genes in coupling and repulsion
In diagrams A–D only one chromatid of each pair is shown
(b)   Diagram illustrating how chiasma-formation leads to breakage of linkage or 'crossing-over'
(c)   Formation of double cross-overs as a result of two chiasmata forming on one bivalent
In this case, for simplicity, only the two chromatids involved in the exchange of material are represented

genes shown in coupling (Figure 22(a), A) recombine, the new arrangement is C. If genes in repulsion, B, recombine, the result is shown by D. Note that the parental types of A and B are the recombinant types of C and D.

Crossing-over

The physical basis for this exchange of allelic genes, or *crossing-over*, between homologous chromosomes is the process of chiasma-formation during prophase I of meiosis.

Suppose that the genes determining purple flowers and long pollen are situated on either side of a chiasma as in Figure 22(b), E. Breakage and re-joining of the chromatids to their opposite partners will produce recombination, F. At the completion of meiosis, such a cell will produce two gametes with the parental combination of these genes and two with the recombinational arrangement. Since the majority of cells undergoing meiosis may not form a chiasma between these gene-sites (or *loci*) and will therefore produce only parental-type gametes, the predominance of the parental type over the recombinational type is explained.

It should now be clear how remarkable were the results of Mendel's experiments. Pea plants have a haploid number of seven, that is, seven different linkage groups. One gene in each of these groups is the maximum number that could possibly segregate independently. We shall never know how Mendel decided to use an appropriate seven pairs of alleles, one from each linkage group.

CHROMOSOME MAPPING

Genes are thought to occur in linear order along the chromosomes with alleles occupying corresponding positions in the homologues. The greater the distance between linked genes, the greater will be the probability of their crossing-over. The fact is used to *map* chromosomes, or to plot the distance between their genes. This technique was developed by T. H. Morgan's famous research group at Columbia University in the United States. The units are arbitrary, one *map-unit* is defined as the distance separating genes between which one per cent crossing-over occurs. A complicated procedure is needed to determine deviations from a 9:3:3:1 ratio as a measure of linkage, so data are accumulated from the results of test-crosses. If an organism, heterozygous for two pairs of alleles, is test-crossed with a double recessive, the ratio between the four possible offspring phenotypes is 1:1:1:1 if there is no linkage. Deviations from this ratio indicate linkage and are simple to handle.

As an example of the kind of experiment Morgan performed, consider data from a cross between a female wild-type *Drosophila* pure-breeding for both red eyes and normal wings and a male with cinnabar eyes and vestigial wings. Letters representing the unusual (or *mutant*) forms are used, with the addition of $^+$ when referring to the wild-type. ($^+$ alone may be used when there is no ambiguity.) When dealing with cases of linkage it is conventional to represent the genes above, rather than alongside, their alleles, to symbolise their arrangement on paired chromosomes.

Parents: ♀   ×   ♂

$$\frac{cn^+\ vg^+}{cn^+\ vg^+} \qquad \frac{cn\ vg}{cn\ vg}$$

red eyes   cinnabar eyes
normal wings   vestigial wings

F_1:   All $\dfrac{cn^+\ vg^+}{cn\ vg}$

red eyes
normal wings

Test-cross A: ♀   ♂   ♀   ♂

$$\frac{cn^+\ vg^+}{cn\ vg} \times \frac{cn\ vg}{cn\ vg} \qquad \begin{array}{c}\text{reciprocal}\\ \text{test-cross B}\end{array} \qquad \frac{cn\ vg}{cn\ vg} \times \frac{cn^+\ vg^+}{cn\ vg}$$

Table 3   Results of test-crosses

	Phenotypes			
	Cinnabar eye, vestigial wing	Cinnabar eye, normal wing	Red eye, vestigial wing	Red eye, normal wing
Test-cross A:	84	14	10	92
Test-cross B:	78	0	0	95

The proportions of the four different phenotypes are assumed to be a direct reflection of the proportions of the four kinds of gametes from the heterozygote. With other organisms, the result of the two reciprocal test-crosses would be similar but male *Drosophila* are peculiar in that they never exhibit crossing-over. Considering crossing-over only in the female, then, we can say that

$$\frac{(14+10)}{(84+14+10+92)} \times 100\% \quad \text{i.e.} \quad 12\%$$

crossing-over had occurred. This is called the *cross-over value (c.o.v)*. The gene loci for cinnabar/red eye and vestigial/long wing are situated on the same chromosome, twelve map-units apart.

If the distance between the genes cinnabar eye and 'humpy' backed (hy) is calculated now by the same method, it is found to be thirty-six map-units.

There are therefore two possible arrangements for these three genes:

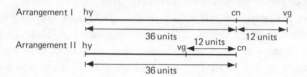

A further cross between hy and vg is necessary to decide which arrangement is correct. A cross-over value of twenty-four per cent between these two establishes that it is arrangement II.

Usually, the quicker method of test-crossing three linked genes simultaneously is used. This is called a *three-point test-cross*. Allowance must be made for the occurrence of *double cross-overs*, (see Figure 22(c)). Chromosome maps have been worked out in detail for certain species.

SEX DETERMINATION

In a large number of organisms, including human beings and *Drosophila*, the female sex-chromosomes are identical, rod-like X-chromosomes. The male possesses one rod-like X-chromosome and one (usually smaller) Y-chromosome, which is hooked in *Drosophila*. A female's eggs will each contain one X-chromosome: she is said to belong to the *homogametic* sex. The male will produce two kinds of sperm in equal proportions: one half containing X-chromosomes and the other half Y. Because there are two sorts of gametes the male is called the *heterogametic* sex.

When the gametes join at random during fertilisation, half the zygotes will receive two X-chromosomes which will determine that they are female, while the other half, receiving one X-and one Y-chromosome, will be male.

The male is not always the heterogametic sex. In birds the female is XY and the male XX. In certain insects the female is XX and the male XO, the Y-chromosome being absent.

SEX LINKAGE

Other experiments in Morgan's laboratory investigated the inheritance pattern of white eye-colour in *Drosophila*. White-eyed males mated with pure-breeding, wild-type females with red eyes gave all red-eyed F_1 offspring. When these selfed to produce an F_2, the ratio of red:white was a perfectly conventional 3:1 – conventional, that is, except for the fact that every white-eyed fly was male. When the reciprocal cross was performed between a red-eyed male and a white-eyed female, half the F_1 were red-eyed females and the other half were white-eyed males. Selfed, this F_1 gave four F_2 phenotypes in equal proportions: one white-eyed female: one white-eyed male: one red-eyed female: one red-eyed male. Obviously, the sex of the fly was connected in some way with the colour of its eyes; this was the first recorded example of *sex-linked inheritance* or *sex-linkage*. (Linkage not involving sex is called *autosomal linkage*.)

The explanation is simple. The gene determining white eye-colour is present on the X-chromosome and, since white eye is recessive to red, the female (XX) must be homozygous in order to show the white-eye character. The male's Y-chromosome is largely genetically inert and so does not carry a normal gene capable of masking the effects of a white-eye gene on his X-chromosome.

Morgan's crosses are represented diagrammatically as shown in Figure 23 using the symbol / for an X-chromosome and $^\wedge$ for a Y-chromosome.

This appearance of a character in father and grandson, apparently 'skipping' the daughter in between is characteristic of sex linkage.

Well-known patterns of sex-linkage occur in human beings. The gene responsible for red-green colour-blindness is carried on the human X-chromosome. This form of colour-blindness is therefore much more common in men, who need only one colour-blindness gene to suffer from its effects, than in

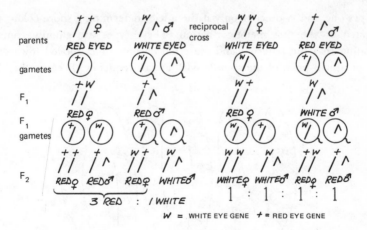

Figure 23   A cross between a red-eyed female *Drosophila* and a white-eyed male, and the reciprocal cross between a white-eyed female and a red-eyed male. These eye-colour alleles are sex-linked, that is, carried on the X-chromosome

women, who need two. Haemophilia, the disease in which blood fails to clot after even minor injuries, is another example whose inheritance follows the same pattern. There is some evidence that a change from the normal gene to the haemophilia gene (such a change is called a *gene mutation*) occurred in Queen Victoria, who subsequently bore a haemophiliac son and transmitted the gene through her daughters.

LINEAR TETRADS

In performing a test-cross between two diploid organisms, one heterozygous and one doubly recessive for the alleles in question, we assume that the proportions of the offspring reflect the proportions of the heterozygote's gametes. The gamete frequencies are inferred from the phenotypic frequencies. A great deal of elegant genetical research has centred on the ascomycete fungi, particularly *Neurospora* and *Sordaria*, because they allow analysis of gametes directly and they are haploid so that recessive mutations are immediately detected.

The fungal body is a dense tangle of threads called a *mycelium*. Meiosis, immediately following the fusion of two mycelia, gives rise to linear *tetrads* (groups of four) of spores which are confined within narrow sacs called *asci*. Their sequence is fixed by the arrangement of the chromosomes and chromatids during meiosis. Eight spores are found in four identical pairs in each ascus, because the four nuclei later divide mitotically.

Consider a cross between a strain of *Neurospora* possessing black spores and one possessing white spores. The convenience of using genes which affect spore colour is that the spores do not have to be grown in order to discover their phenotype. The heterozygous individual, Aa, produces six different kinds of tetrad (see Figure 24).

Types one and two occur only if the alleles determining spore colour were separated at the first meiotic division. These types are equally common, indicating that the paternal and maternal chromosomes moved towards the base, or apical pole, of the spindle with equal frequency. Because the alleles

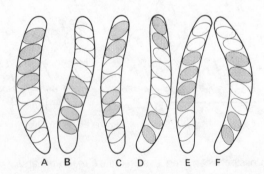

Figure 24 The six possible arrangements of spores within the asci of the ascomycete fungi such as *Neurospora*

separate at the first meiotic division these arrangements are called *pre-reductions*. Types 3, 4, 5 and 6 indicate that the alleles separated at the second meiotic division and, therefore, crossing-over has occurred between the centromere and the gene locus in question. Again, the four types appear with equal frequency, indicating that there is no preference in the orientation of the dividing chromatids. These types are called *post-reductions*. Figure 25 clarifies the relationship between the behaviour of the chromosomes and the spore arrangements in the asci.

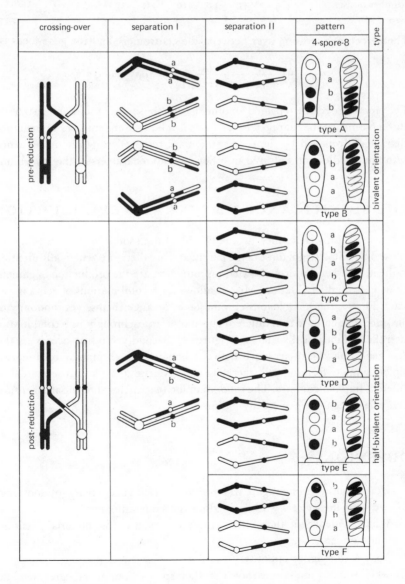

Figure 25   The relationship between chromosome behaviour and spore pattern in the linear tetrads of the ascomycetes

The relative frequency of the pre- and post-reductions depends upon the frequency of crossing-over between the centromere and the locus. As in the construction of *Drosophila* chromosome maps, the frequency of crossing-over is used as a unit of map-distance. For example, suppose that a collection of asci were counted into types as follows:

Table 4

Type	1	2	3	4	5	6
Number of asci	54	57	12	10	11	14

The percentage crossing-over between the centromere and the gene-locus is:

$$\frac{(12+10+11+14)}{(54+57+12+10+11+14)} \times 100\% = 29.7\%$$

The procedure can be repeated for other loci on the same chromosome so that the chromosome can be mapped. Since there is a limit to the number of structural characters displayed by the spores, these are removed carefully, in sequence, with fine needles and cultured in order to study characters of the mycelium.

THE SIGNIFICANCE OF SEXUAL REPRODUCTION

When organisms reproduce asexually, the chromosomes of the offspring are derived by mitosis from those of the parent and are therefore genetically identical to them. The key features of sexual reproduction are the production of gametes by meiosis, and fertilisation. Meiosis allows for recombinations of genes in two ways. Firstly, the homologous chromosomes arrange themselves randomly on the spindle equator so that different chromosomes are brought into combination after the first meiotic division (see Figure 21). Secondly, crossing-over is usual so that genes may be re-arranged between homologous chromosomes. Identical diploid gamete-forming cells therefore can give rise to an immense number of different haploid gametes. The random union of gametes at fertilisation ensures that every zygote is genetically unique.

QUESTIONS

1 How would you investigate whether the mutant genes ebony (e) and rough eyes (ro) in *Drosophila* are located on the same autosome?
2 What evidence is there that chromosomes are the material basis for inheritance?
3 Look at the figure opposite:
   (a) How many ascospores show that their spore colour genes segregated after the first meiotic division? How many segregated after the second division?
   (b) Calculate the percentage of crossing-over.

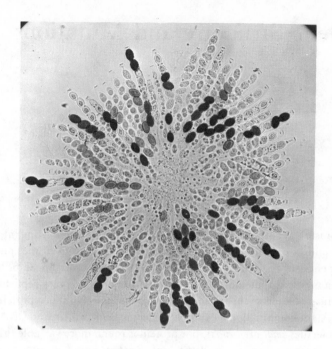

4 Counts were made of the number of cells in different stages of mitosis in an onion root-tip.

Stage	Number of cells
Prophase	623
Metaphase	95
Anaphase	35
Telophase	148

What percentage of the total time for one mitotic division does each stage last?

5 Explain the significance of meiosis and fertilisation in producing a range of genetic variation in offspring.

6 Yellow female cats crossed with black toms produced tortoise-shell female kittens and yellow male kittens. Black females crossed with yellow toms produced tortoise-shell females and black males. In each case, male and female kittens were produced in approximately equal numbers. How would you explain this pattern of inheritance? Draw diagrams to illustrate the crosses. What offspring would result from crossing:
(a) two black cats
(b) two yellow cats?

4 Gene Structure and Mutations

THE GENETIC MATERIAL

An organism's genotype is composed of a collection of genes storing information which, under certain conditions, can be translated into the phenotype. Experiments suggest that genes occur at specific loci in linear order along chromosomes and that they replicate when the chromosomes replicate.

Chemical analysis of chromosomes reveals that they contain a protein, *histone*, and DNA; the combination is called *nucleoprotein*. Both protein and DNA are very large molecules (*macromolecules*), but protein is much more variable than DNA. For a long time it was assumed that only protein had a structure capable of sufficient variety to carry all the coded information necessary for the development of a living organism.

Bacterial transformation

In the 1920s, work began on the pneumococcus bacterium *Diplococcus pneumoniae*, which causes pneumonia in mice; the work eventually led to the identification of the genetic material. Two distinct forms of pneumococcus occur, one of which secretes a *polysaccharide* (complex carbohydrate) capsule, which gives the colonies a smooth appearance, and another which has no capsule and produces small rough colonies. The capsule of the smooth form (S) gives an infected animal the disease pneumonia (it is said to be *virulent*); the rough form (R) is non-virulent. The presence or absence of a capsule is genetically determined and different types of capsule (I, II, III, IV) with specific inheritable characteristics are recognised. About one S bacterium in every 10^7 mutates (changes genetically) into an R form. After many generations, a descendant of this R bacterium may mutate back to form an S and, if it does so, it always produces the same type of capsule as the one from which the R colony was originally derived.

Mice were injected with live non-virulent type-IIR bacteria and showed no ill-effects. Others, injected with heat-killed virulent type-IIIS bacteria, survived equally well. When both live IIR and dead IIIS bacteria were simultaneously injected into the same mice, however, the mice died with the symptoms of pneumonia. Living type-IIIS bacteria were recovered from their bodies; this ruled out the possibility that IIR bacteria mutated into a virulent smooth strain, as they would inevitably mutate to the IIS type. This remarkable discovery had two possible explanations: either the dead bacteria had been resurrected by the live ones or else something from the dead bacteria had entered the live ones and made them virulent. Live IIR bacteria plus extracts of type-IIIS in the absence of the whole cell also killed mice, so proving that the second possibility was the correct explanation. This phenomenon is called bacterial *transformation*: something from the dead bacteria is acquired by the living bacteria and carries to them genetic information for a different phenotype. The next question was: 'What *is* the transforming substance?'.

Identification of the transforming substance

In 1944, O. T. Avery, C. M. MacLeod and M. McCarty conducted a series of painstaking experiments in which they tested heat-killed fractions of S bacterial extracts for their transforming ability. One extract contained only polysaccharide capsules, another just purified *RNA* (*ribonucleic acid*, a substance closely related to DNA), a third contained DNA alone. Many different extracts were tried and each tube was seeded with a small inoculum of active R-bacteria. Only the DNA extract plus live R-bacteria proved capable of infecting mice with pneumonia. This showed that DNA was the transforming substance which specified that the living bacteria should produce polysaccharide capsules and hence become infective. The conclusion seemed definite enough, and yet it was not immediately widely accepted. Protein had long been regarded as the obvious genetic material and biologists were reluctant to accept that this rôle belonged to DNA.

The following decade saw increasing interest in nucleic acids and the work of Avery, MacLeod and McCarty was repeated and confirmed. Bacterial transformation was found to be a widespread phenomenon which occurred even between different species of bacteria. Pneumococci with a resistance to the antibiotic streptomycin were found to be capable of transferring this resistance to streptococci (a different kind of bacterium). This is a serious problem in the treatment of disease since drug-resistance can spread rapidly throughout bacteria by transformation much faster than bacteria can acquire resistance by mutation.

Genetic material in bacteriophage

Work by Hershey and Chase in 1952 established that DNA, not protein, carried the genetic information in *bacteriophages* (viruses which attack bacteria). Viruses consist of a protein shell with an inner core of nucleic acid (Figure 26). The well-studied T_4 bacteriophage (*phage* for short) attaches itself to the outside of a host bacterium by means of its tail fibres, injects a substance into the bacterium and, within half-an-hour, the bacterium bursts, releasing several hundreds of fully-formed phage particles. Whatever is injected must carry genetic information which dictates the rearrangement of the host-bacterial cell to form new phage.

Hershey and Chase inoculated *Escherichia coli* (the colon bacterium) on to a medium containing the radiocative isotope of sulphur, ^{35}S. Another inoculation of *E. coli* was made on to a different medium containing radioactive phosphate, ^{32}P. As the bacteria multiplied, they incorporated their respective isotopes into

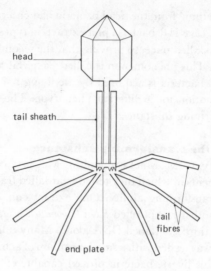

head

tail sheath

end plate

tail fibres

Figure 26 A bacteriophage particle

their bodies. The bacteria were then infected with phage. Since sulphur is a constituent part of protein but not of DNA, the phage progeny collected from the ³⁵S-fed bacteria contained radioactively labelled protein. Conversely, phosphate occurs in DNA but not in protein, so the progeny from the ³²P bacteria had radioactive DNA. The two types of labelled phage progeny were collected by centrifugation and allowed to infect two separate batches of unlabelled host bacteria. The phage and bacteria were then separated and tested for radioactivity. It was found that the phosphate label was associated with the bacterial cells but the sulphur label was found in the empty phage coats which washed off the bacteria. Evidently, the DNA had penetrated the bacteria but the protein coat had not. Only the DNA conveyed the information for the formation of the next generation.

Similar evidence showed that RNA is the genetic material in tobacco-mosaic virus, and in 1956 Fraenkel-Conrat and Williams reconstituted infective virus particles from the RNA of one strain and the protein of another. The two strains differ in the kind of lesions they produce on the tobacco leaves and the experiments demonstrated that the lesions corresponded to those produced by the strain which provided the RNA, not that which provided the protein. Clearly, nucleic acid determined the characteristics.

DNA: the genetic material

Although the most direct evidence comes from a study of microbes, biologists now agree that DNA is the nearly-universal genetic material. Only certain viruses use RNA rather than DNA.

A genetic material which carries encoded material from generation to generation must not be metabolised. Measurements of the quantity of DNA in each cell show that the amount is constant for different somatic cells in one

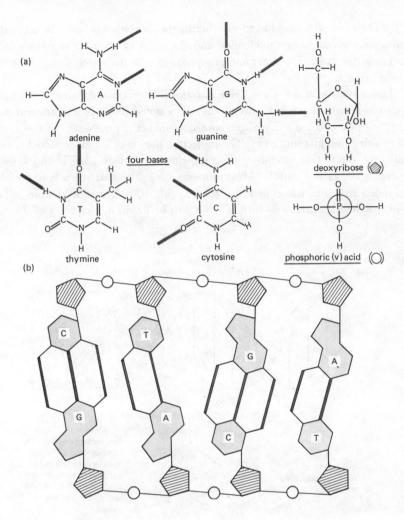

Figure 27(a)   Components of DNA
The shaded bars show the positions at which the bases form hydrogen bonds
(b)   DNA molecule
The backbone of the molecule is made from deoxyribose and phosphate units, while the bases pair A–T and G–C across the middle

organism except during chromosome duplication, when the amount doubles. Haploid gametes of diploid organisms contain half the somatic quantity of DNA per cell.

The structure of DNA

DNA is made up of units called *nucleotides*, each comprising a phosphate group, a five carbon sugar called deoxyribose and an organic *base* which is either a *purine* or a *pyrimidine* (Figure 27). There are two double-ring purine bases called adenine (A) and guanine (G) and two single-ring pyrimidines, cytosine (C) and thymine

(T). The nucleotides are linked together by the formation of bonds between the phosphate group of one nucleotide and the sugar of the next to form a long *polynucleotide* chain. The sugar and phosphate form a 'backbone' from which the bases project.

James Watson and Francis Crick at Cambridge proposed a structure for the DNA molecule shown in Figure 28 (a). They noted that a pair of adenine and thymine bases, joined by hydrogen bonds, would occupy the same space as a pair of cytosine and guanine bases. The molecule, they concluded, consisted of two adjacent nucleotide strands with complementary bases $(A+T$ and $C+G)$ meeting across the middle. Measurements suggested that the whole double-stranded structure was coiled into a helix, that is, it formed a 'double-helix'. Further details of the structure of DNA can be found in chapter 3 of *The cell concept*.

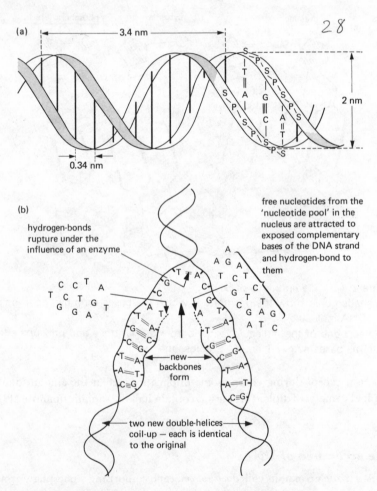

Figure 28(a)   The double-helix structure of a molecule of DNA
S= sugar; P= phosphate; A= adenine; T= thymine; C= cytosine; G=guanine
(b)   Replication of DNA

The specific base-pairing of the DNA molecule suggested to Watson and Crick an accurate copying-mechanism by which it could duplicate itself. If the relatively weak hydrogen-bonds holding the bases together were ruptured, the exposed bases would be available to join with other similar ones from a pool of unattached nucleotides in the cell. Each helix could build itself a new partner by acting as a template so that two double helices would be derived from the original (Figure 28 (b)).

Watson and Crick's model fulfilled two of the basic requisites for a structure bearing the genetic code: it was capable of self-replication and it possessed tremendous potential for variety. Since any base may follow any other base along the length of one helical chain, the number of possible arrangements is 4^n where n is the number of base pairs in the double helix.

Watson and Crick published their theory in a letter to the scientific journal *Nature* in 1953. In 1962, after their proposal had been substantiated, they received shares in a Nobel prize for their work.

Replication of DNA

Watson and Crick had envisaged that the DNA molecule would replicate by splitting longitudinally so that each chain acquired a new complementary partner. Each new molecule, therefore, would consist of one 'old' and one 'new' DNA strand. This mechanism is termed *semi-conservative replication*. If the whole 'old' double-helix remained intact and a new one assembled alongside, this would be *conservative replication*.

An experiment by Meselson and Stahl in 1958 provided very good evidence that replication is indeed semi-conservative (Figure 29). Bacteria were allowed

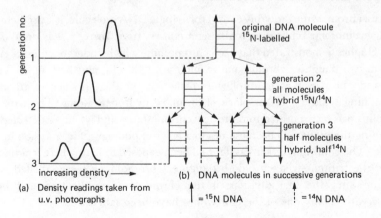

(a) Density readings taken from u.v. photographs

(b) DNA molecules in successive generations

$\uparrow = ^{15}N$ DNA $\uparrow = ^{14}N$ DNA

Figure 29   Diagram of semi-conservative replication of DNA, as demonstrated by Meselson and Stahl's experiment

Bacterial DNA molecules, all labelled with 'heavy' ^{15}N, are allowed to replicate on ^{14}N medium. The next generation of DNA molecules are uniformly less dense than those of the original generation. By the second generation of replication on light medium, equal proportions of 'light' and 'hybrid' DNA are present to grow on a medium in which the nitrogen source was labelled with the heavy isotope, nitrogen-15 (^{15}N). After many generations, virtually all their DNA bases were labelled. The bacteria were then changed to a medium with the usual ^{14}N. After one cell-division, each DNA double-helix should contain one ^{15}N strand and one ^{14}N strand if the replication is semi-conservative, but strands would be all ^{15}N or all ^{14}N if replication were conservative. To find out which was true, Meselson and Stahl relied upon the fact that the heavy isotope produces a slightly denser DNA. Their technique for measuring subtle density differences entailed centrifuging a solution of caesium chloride at high speeds. Caesium chloride is a dense substance and under centrifugal forces of over 100000g the solution becomes denser at the bottom of the tube than at the top. If DNA is suspended in this density gradient it will find its own level. Meselson and Stahl's technique involved extracting DNA from bacteria and photographing the position of the band of DNA with ultra-violet light through quartz windows in the running centrifuge. (DNA absorbs ultra-violet light strongly.)

The experiments showed that, after one division, the DNA occupied an intermediate position in the gradient between that of all-heavy DNA and all-light DNA. After two cell divisions on ^{14}N medium, the DNA showed two bands, one in the 'intermediate' position and one in the 'light' position. Subsequent divisions showed increases in the proportion of 'light' DNA. These results are consistent with the model of semi-conservative replication. Further proof was furnished when the 'intermediate' DNA was boiled, a treatment which separates the two helices. On centrifuging in the gradient, one 'light' and one 'heavy' line were observed.

Duplication of chromosomes

Since a chromosome probably consists of a single DNA molecule, it is predictable that chromosome replication is also semi-conservative. In 1957, Taylor, Woods and Hughes demonstrated that each chromosome which undergoes mitosis gives rise to two daughter chromosomes which are half 'old' and half 'new'. They allowed chromosomes to duplicate themselves in the presence of thymine containing tritium (^{3}H) in place of ordinary hydrogen atoms. The tritiated thymine was incorporated into the new chromosomes and was later detected by exposing the prepared chromosomes to a layer of photographic emulsion in the dark. (The emulsion darkens as if it had been exposed to light where irradiation from the tritium strikes it.) Both daughter chromosomes were labelled with tritium after the first division; if the chromosome duplication had been conservative, only one of the pair would have been labelled.

MUTATION

A particular length of DNA (usually 600–1800 nucleotide pairs) constitutes one gene. Genes occasionally undergo changes called *mutations* which are inheritable. Mutations are random in their occurrence and effect and, since it is rare that a gene's delicately-balanced function is improved by such a change, most mutations are deleterious. In a diploid organism, the relationship between the mutant gene and the normal 'wild-type' gene is important. Since mutation is an infrequent event, most organisms possessing a newly-arisen mutant gene will be heterozygous for that gene. If the mutant's effects are dominant to those of the wild-type gene, the organism is usually at a disadvantage compared with the normal type and is eliminated by natural selection. A mutant gene with recessive effects, however, produces no apparent change in the organism's phenotype and may be transmitted to the offspring. Its effects show only when a doubly-recessive homozygote arises from a mating of heterozygotes.

While the occurrence of a particular mutation cannot be predicated, genes have their own characteristic rate of mutation. Bacterial genes mutate with a frequency of approximately 1 in 10^7 and those of fruit-flies between 1 in 20000 and 1 in 200000.

Such changes in the DNA of a single gene are called *point-mutations* to distinguish them from aberrations of whole or part chromosomes which will be considered later. Evolution depends on mutations for providing the source of variety on which natural selection can act. It is worth pondering how this can be so if most mutants are harmful and must be recessive in order to persist. It must be remembered that neither 'harmfulness' nor 'recessiveness' are fixed properties of genes.

Mechanism of point-mutation

The smallest possible point-mutation is a change in one base-pair and this occurs if a base alters in such a way that it pairs with the 'wrong' partner, for instance, thymine partners guanine and cytosine partners adenine. Certain chemicals can induce mutation by altering DNA bases or by being structurally so similar to normal bases that they are incorporated into a replicating DNA strand where they cause wrong pairing.

Mutagens

Any substance which increases the spontaneous mutation rate is termed a *mutagen*. The ability of X-rays to act as mutagens of fruit-fly genes was discovered by H. J. Müller in 1927. Since then, other types of ionising radiation such as α, β and γ rays have been found to have the same effect.

In the case of point-mutations, there is a linear relationship between the dose of radiation received and the rate of occurrence of mutation. The effect is cumulative, so that an X-ray dose of 8000 röentgen units, R, (the measure of the amount of ionisation produced) administered at one time causes the same percentage mutation as four 2000 R doses applied separately. This relationship applies more exactly to gametes than active somatic cells, because the repair systems operating in active cells can make good some of the damage caused by regular low-dose irradiation which is, as a result, less harmful than one equivalent amount of radiation in a single large dose. Chromosomal breakage can also be induced by irradiation.

Ultra-violet light, though less penetrating than X-rays, is strongly absorbed by DNA and has mutagenic properties. The number of recognised mutagenic chemicals increases annually; it includes mustard-gas, nitrous acid, phenols, hydrogen-peroxide, formaldehyde and caffeine.

Radiation damage to the somatic tissue affects only the individual concerned, but damage to testis or ovary cells (*germ cells*), which proliferate and then divide to form sperm or eggs, is genetically more serious. An irradiated germ cell may give rise to many gametes, each of which carries mutations. The higher the proportion of damaged gametes, the greater is the chance that one of them will be involved in fertilisation. The resulting zygote, if viable, will undergo many cell divisions to produce an embryo in which every cell carries a set of chromosomes with the same mutations. The effects on the individual's phenotype may be profound. If mutation occurs during the rapid cell multiplication of embryonic growth only a proportion of cells will be affected. The proportion of cells affected depends on the stage of development at which mutation occurs. (Dividing cells are particularly susceptible to the effects of mutagens, but mutation can happen at any time in the cell cycle.)

Multiple alleles

A mutation which is not eliminated by natural selection will persist in the genotype of the descendants of the original mutant organism. There will then be two alleles of the same gene in the population. Genes at the same locus may mutate in more than one way so that, eventually, a number of alternative forms of one gene occurs. These are called multiple alleles. A diploid organism can, of course, possess only two of the possibilities.

One series of multiple alleles is that which determines human blood groups. Red blood cells may contain either or both of two sorts of *antigens* (complex proteins) called A and B. A-group blood contains antigen A, B-group contains antigen B, AB-group contains both antigens and O-group contains neither antigen. *Antibodies* are large protein molecules which combine with a particular antigen, so causing the red cells to clump together. Thus antibody (or anti-) A in the blood of a B-group person will clump A-blood and anti-B in A-group blood will clump B-blood. Ignorance of these facts accounted for many fatalities in the transfusion of blood before the discovery of blood-groups by Landsteiner in 1900.

The genotype of group A-blood is either $I^A I^A$ (homozygous) or $I^A i$ (heterozygous). B-group blood is determined by the allele I^B so that the possible genotypes are $I^B I^B$ and $I^B i$. The homozygous recessive genotype, ii, represents O-blood. Alleles I^A and I^B are *co-dominant*, that is, when they are present together they both produce their effects so that $I^A I^B$ is the genotype of AB-blood. Both I^A and I^B are dominant to i. This multiple allele series has been stable for a long time and it is thought that the different blood groups are associated with susceptibilities to certain diseases.

Chromosome mutations

Mutations are heritable alterations of genotype, a definition which can extend to changes in whole or part chromosomes. Because many genes will be affected simultaneously by a chromosomal mutation, these often have drastic effects.

Chromosome mutations can be studied cytologically in great detail, thanks to the discovery of giant chromosomes in the salivary glands of certain insects. These *polytene*, or many-stranded, chromosomes are the product of many duplications of the chromosomal material without the normal subsequent separation. The result is chromosomes which are at least 100 times the size of normal chromosomes undergoing meiosis.

Each chromosome has a pattern of conspicuous transverse bands which are constant in size and spacing for any particular normal chromosome. These have been mapped out with great accuracy in *Drosophila*, whose giant chromosomes are particularly favourable for study. Any chromosome mutation can be examined and related to alterations in the phenotype revealed by breeding experiments.

Each species of organism is characterised by a particular complement of chromosomes (the *genome*) which is represented once in haploid cells and twice in diploid cells. Chromosomal mutations may involve the loss or addition of whole genomes or single chromosomes. Structural changes occur if chromosomes break and the broken parts reunite in a different arrangement. Such modifications may involve only a portion of one chromosome.

Structural changes (see Figure 30)

1 Deficiency (Figure 30(a))
A segment of a chromosome may become lost, a case known as a *deficiency*. At pairing during meiosis, the homologue of the deficient chromosome will form a characteristic loop which allows the transverse bands to match as well as possible. The human 'cri du chat' syndrome, with symptoms of severe mental retardation, a round moon-face and a cat-like cry, is the result of a deficiency on chromosome number 5.

2 Duplication (Figure 30(b))
A fragment of chromosome may become duplicated so that two identical segments are joined end to end in place of one. There are, therefore, extra 'doses' of those genes contained in the duplicate segment. A well-researched example is that of 'bar-eye' in *Drosophila*, in which a repeated segment of the X-chromosome has the effect of reducing the number of eye-facets. 'Double-bar' individuals are known, in which a further duplication gives three identical segments. Individuals which are homozygous for bar-eye and therefore have four sets of 'bar' genes instead of the normal two, have been produced by breeding experiments. They average sixty-eight facets per eye. Flies which are heterozygous for double-bar also carry four sets of bar genes, but this time three of these are on one chromosome and one is on its homologue. Such flies average only forty-five facets per eye, indicating that the three adjacent genes reinforce one another more strongly than two pairs do. The discovery of this *position effect* was the first

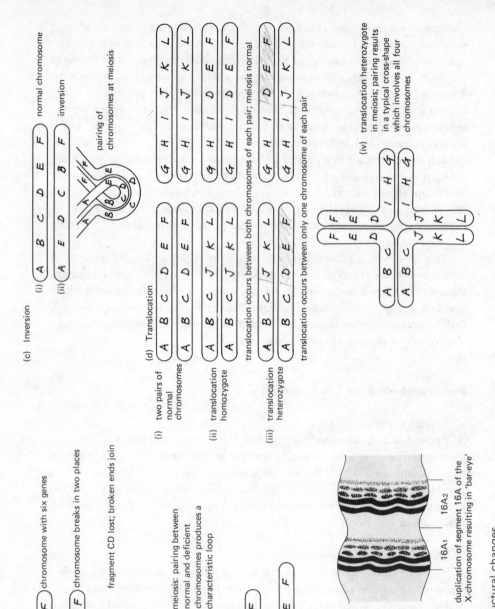

Figure 30 Chromosome structural changes

indication that the position of a gene on a chromosome (and hence its genetic neighbours) was important in the expression of the gene.

3   Inversion (Figure 30(c))

Inversions occur when portions of chromosomes become detached and rejoin so that the genes are in reverse order. Pairing at meiosis is accompanied by loops in both the inverted segment and its homologue.

4   Translocation (Figure 30(d))

Translocations involve fragments of one chromosome being transferred to a non-homologue. If parts are exchanged, the result is a *reciprocal translocation*. *Translocation homozygotes* are those in which both members of a homologous pair of chromosomes exchange similar sections with both members of a different homologous pair. After the double exchange, both sets of homologues are capable of normal meiosis. *Translocation heterozygotes* are those in which only one chromosome of each pair exchanges a segment so that there is difficulty in attaining a matching of homologous parts during meiosis. A cross-shaped formation is typical of prophase-I which may open out into a ring as the chiasmata terminalise. Many gametes fail to receive the full complement of genes so that partial sterility can result.

Changes in chromosome number

1   Euploidy

Euploidy is the possession of complete sets of chromosomes and aberrants occur which have extra active sets, a condition known as *polyploidy*. Triploids may occur in plants if an egg's chromosomes fail to separate during meiosis so that a diploid egg is fertilised by a haploid pollen-grain to give three genomes. The plant may grow successfully and propagate vegetatively, but sexual reproduction is not possible because the chromosomes cannot all pair during meiosis. Tetraploidy, or the possession of four genomes, occurs if chromosome duplication in a normal zygote is not followed by cell-division, or if two diploid gametes fuse. In this case, chromosome pairing is possible and meiosis may be successful. Higher numbers of genomes occur regularly in plants: the rose genus *Rosa* includes species with 14, 21, 28, 35, 42 and 56 chromosomes in the somatic cells. This probably represents a series of polyploids which originated from a diploid with 14 chromosomes per cell. Tetraploids in which all four genomes originate from one species are called *autotetraploids*.

A cross-fertilisation between the species of primula called *Primula verticillata* ($2n = 18$) and *P. floribunda* ($2n = 18$) at Kew in 1900 resulted in a hybrid that was named *P. kewensis*. This also had a diploid number of 18, but the plant was sterile since the two sets of 9 chromosomes were not homologous and so could not pair during meiosis. After five years of vegetative cultivation, a fertile shoot appeared which had a diploid number of 36. Such chromosome doubling in a diploid hybrid is termed *allotetraploidy* (the general case is known as allopolyploidy). It is an abrupt method of species-formation, since the new tetraploid cannot cross with the original diploids. A similar series of events is thought to have caused the evolution of the marsh grass *Spartina townsendii* ($2n = 126$) from the hybridisation of *S. alterniflora* ($2n = 70$) and *S. stricta* ($2n = 56$). The original diploid hybrid ($2n = 63$) may have produced some unreduced gametes ($n = 63$) which self-fertilised to give the fertile autotetraploid known today. *Spartina townsendii*'s rapid spread along Channel coasts is probably due to its hardiness and vigour, characteristics of polyploids which account for the polyploid nature of many domestic plants such as bread-wheat, cotton and melon.

Polyploidy is virtually confined to plants since, as well as the ability to self-fertilise, they can often reproduce vegetatively, which provides favourable conditions for the rare chromosomal doubling to occur. Also, animals possess a sex-chromosome mechanism which would render polyploidy non-functional.

2 Aneuploidy

Aneuploids have either one chromosome too many or one too few in the nuclei of their cells. The possession of an extra chromosome ($2n + 1$) in the diploid cells is called *trisomy* and the lack of one chromosome ($2n - 1$) is called *monosomy*.

Aneuploidy usually arises as a result of *non-disjunction*, or unequal separation of chromosomes during meiosis, so that one gamete receives two homologous chromosomes and the other none. On fertilisation of the gamete with an extra chromosome, a trisomic zygote is formed, while the fertilisation of the gamete lacking a chromosome results in a monosomic zygote.

The imbalance of genes resulting from aneuploidy often has drastic effects so that the organism dies. In animals, aberrations in the number of sex-chromosomes is less frequently fatal than alterations in somatic chromsome number. *Drosophila* eggs, with two X-chromosomes, on being fertilised by normal sperm, produce XXX metafemales, which are sterile, and XXY individuals, which appear to be normal fertile females.

As many as one in 200 human babies may be born with aneuploid cells, about half of these having aberrations of the sex-chromosome number. Turner's syndrome results from a loss of a sex-chromosome, XO. The individual appears to be female, but the ovaries do not develop, the body is short, the neck has a fold or web of skin and there are defects in the jaw and chest. An abnormal male condition called Klinefelter's syndrome is trisomic, XXY. The testes do not develop fully and there is a tendency towards the development of female secondary sexual characters and mental retardation.

The best-known autosomal trisomy causes *Down's syndrome* previously called 'mongolism'. Chromosome number 21 is represented three times in such individuals (Figure 31). The syndrome includes mental retardation, short stature, characteristically slanting eyes, broad hands and heart defects. Older mothers run a greater risk than younger ones of producing a child with Down's syndrome. This may be because human eggs remain quiescent half-way through prophase from before birth until shortly before ovulation. The egg of a woman aged forty has therefore been in prophase twice as long as that of a woman aged twenty so that the chance of non-disjunction is increased.

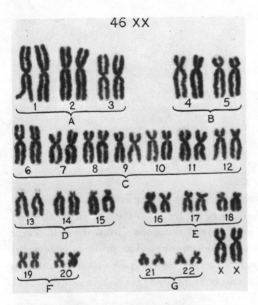

46 XX

Figure 31 (left)   Normal human chromosomes
Figure 31 (below left and right) Photograph and chromosome complement of a person with Down's syndrome

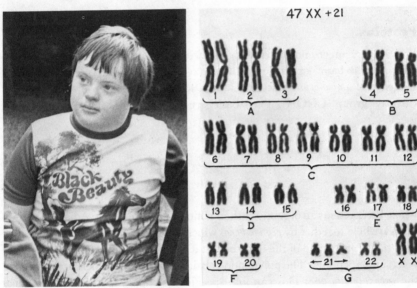

47 XX + 21

QUESTIONS

1   What evidence have we for the statement: 'DNA is the genetic material of all living things'?

2   Tortoise-shell cats are almost always female. An occasional sterile male tortoise-shell arises. Account for these facts.

3   Why is irradiation of the ovary or testis with X-rays considered a greater genetic hazard than the irradiation of other parts of the body?

4   Write an account of chromosome mutations.

5 Gene Action

GENES AND ENZYMES

Genes function by synthesising proteins. Some of these complicated molecules make up the framework of the cells of living tissues, while the majority function as enzymes which control specific biochemical reactions within the cell. These reactions determine the phenotype of the organism. The problem of how the genetic code provides information for the creation of a phenotypic character is really the problem of how the code can be translated into certain proteins. (For details of protein structure, see *The cell concept*.)

Proteins

Proteins are macromolecules with molecular weights ranging from 6000 to 500000. The basic, or *primary*, structure of each molecule consists of a linear sequence of *amino-acids*. These contain at least one amino group (NH_2) and a carboxyl group (**COOH**). Their basic structure is:

$$H-\overset{\overset{\displaystyle H}{|}}{N}-\overset{\overset{\displaystyle H}{|}}{\underset{\underset{\displaystyle R}{|}}{C}}-\overset{\overset{\displaystyle O}{\|}}{C}-OH$$

where R represents an organic (carbon-containing) side-chain. Adjacent amino-acids are held together by *peptide-bonds* which link the carboxyl group of one acid with the amino group of the next to form a *polypeptide*. Large proteins consist of several polypeptides. The polypeptide chain often assumes a helical form called the *secondary structure*. This can fold specifically to give the three-dimensional *tertiary structure*.

About eighty kinds of amino-acid are known, but only twenty of these are commonly found in proteins. The amino-acids can occur in any order in the protein, so the number of possible orders is astronomical, although only a fraction of these possibilities is actually found.

Human metabolic disorders

The relationship between genes and enzymes in humans is illustrated by the inherited disorder *phenylketonuria* (**PKU**) which affects individuals who are doubly recessive for a mutant gene. The normal form of the gene produces an enzyme which converts the amino-acid phenylalanine into the related amino-acid tyrosine. Phenylalanine is an essential amino-acid found almost universally in protein foodstuffs. The mutant form of the gene does not produce the necessary enzyme so that phenylalanine accumulates and is excreted in the urine. The excess phenylalanine results in severe mental and physical retardation. In Britain, the urine of newborn infants is tested for phenylalanine-products and, if the disorder is diagnosed, the child is brought up on a diet containing minimal amounts of this amino-acid. The severity of the symptoms is thereby reduced. In this disease, the absence of a single enzyme has profound effects.

Nutritional mutants in Neurospora

Human metabolic defects are not suitable for the experimental study of the relationship between genes and enzyme-formation. An organism well adapted to such investigations is the red bread-mould, *Neurospora crassa*, originally chosen for study in 1941 by Beadle and Tatum. The mould is normally haploid and produces haploid spores by an asexual process. Under some circumstances, haploid nuclei from two different mating-types may fuse to form a diploid nucleus. This undergoes two meiotic divisions followed by one mitotic division to give eight haploid ascospores, as previously explained in chapter 3. The wild-type *Neurospora* can grow on a 'minimal' culture medium containing basic nutrients which the mould cannot synthesise for itself. Sometimes a mould which has been cultured on a medium containing a wide range of nutrients fails to grow when transferred to minimal medium. This indicates that a mutation has occurred so that some essential biochemical pathway is blocked by the absence of a necessary enzyme. A culture of the mutant strain is then attempted on a variety of minimal media, each of which has a different supplementary substance. If growth occurs only on a minimal medium with, say, additional arginine (an amino-acid), then it is assumed that the mutation has deprived the mould of the ability to manufacture arginine. Beadle and Tatum identified three separate arginine-requiring mutants, however. One can grow only when supplied with arginine, a second can use either arginine or citrulline, while the third utilises arginine, citrulline or ornithine. This discovery implies a sequence of reactions, each catalysed by a separate enzyme:

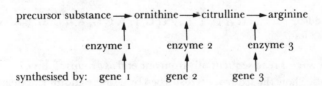

Beadle and Tatum suggested that a single gene was responsible for the synthesis of a particular enzyme in the chain and this proposal became known as the 'one gene – one enzyme' hypothesis. This has since been redesignated the one gene – one polypeptide chain hypothesis, as some enzymes consist of more than one polypeptide, each synthesised by a separate gene.

The idea that genes manufacture enzymes has been substantiated in a variety of different organisms. We must now consider how this is achieved.

THE GENETIC CODE

Most proteins contain several hundred amino-acids and each includes all of the twenty common ones. The order of bases in the DNA of a gene specifies the order of amino-acids in a polypeptide chain, and this in turn determines the folding which gives secondary and tertiary structure to the protein molecule.

The simplest kind of code would be one in which each of the four DNA bases represented one amino-acid but, obviously, twenty amino-acids cannot be encoded in this way. Pairs of bases will not be sufficient either, yielding only 4^2, or 16, combinations. The minimum grouping must be three bases per amino-acid, which allows for 4^3, or 64, possibilities. A group of bases which stands for one amino-acid is called a *codon*.

Direct evidence for this theoretical prediction came from a study of mutations in the T_4 bacteriophage by Crick and his co-workers. Known genes of T_4 are induced to mutate by the application of acridines, compounds which produce additions or deletions of base-pairs. The mutants can be detected by the appearance of circular clear zones, or *plaques*, which appear on the culture plates where the phage particles have multiplied and burst the bacterial cells.

If a bacteriophage carrying a deletion mutation ($-$) is cultured alongside one carrying an addition mutation ($+$), recombinants are produced, some of which bear both mutations. These recombinants often function normally. Thus:

Portion of wild-type gene   G A C, T A G, G T A, C G A, A T C, G T A, A G C,

One base added, $+$,   G A C, T T A, G G T, A C G, A A T, C G T, A A G, C

One base lost, $-$,   G A C, T A G, G T A, C G A, A C G, T A A, G C

One base added, $+$,   G A C, T T A, G G T, A C G, A A C, G T A, A G C,
and one base lost, $-$.

If the message is read sequentially from one end as groups of three bases (called *triplets*), notice how the sequence is disturbed by a single addition or deletion, but with both mutations the message eventually comes in phase again. The distance between the sites of the two mutations determines the number of incorrect triplets that are created, and on this depends the degree of error in the manufacture of that gene's protein.

Powerful evidence for the triplet nature of the code is that while one or two addition mutations upset the code, so that the gene is non-functional, three additions restore function. Similarly, one or two deletions do not allow the gene to work, but three do. Convince yourself of the reason for this.

A triplet code could work either by 'overlapping' or by not overlapping:

Sequence of bases   C A T A T T G A T A G C C T G
Codons of non-overlapping code   C A T, A T T, G A T, A G C, C T G.
Codons of overlapping code   C A T, A T A, T A T, A T T, T T G, T G A, G A T,
A T A, T A G, A G C, G C C, C C T, C T G.
or: C A T, T A T, T T G, G A T, T A G, G C C, C T G.

If the codons overlapped, there would be a restriction of the order of amino-acids in a polypeptide chain (since an amino-acid whose codon was CAT must be followed by one whose codon begins AT or T, depending on the degree of overlap). This is not the case; any amino-acid may be followed by any other amino-acid. Also, a mutation of one base-pair of an overlapping code would alter two or three codons, and point-mutations typically change only a single amino-acid. All the evidence suggests that the code does not overlap.

Reading the genetic code

DNA is found in the chromosomes which are located in the nuclei, yet protein synthesis occurs in the cytoplasm. If radioactive amino-acids are offered to mammalian cells which are synthesising protein, small cytoplasmic structures called *ribosomes* quickly become radioactive, suggesting that they are the sites of the synthesis. Ribosomes consist of protein and *ribosomal-RNA (r-RNA)* forming spheres of about twenty nanometres' diameter.

Two other kinds of RNA are involved in protein synthesis: *messenger-RNA (m-RNA)* and *transfer-RNA (t-RNA)*. Messenger-RNA is synthesised along one strand of an unravelled DNA double-helix in a process called *transcription*. RNA contains similar bases to DNA so base-pairing can occur, but in RNA the base *uracil* pairs with the DNA's adenine in place of thymine (see Figure 32). The information contained in the order of codons in DNA is therefore carried as a complementary codon sequence in the m-RNA. One m-RNA molecule is a copy of a small fraction of the DNA molecule which corresponds to one gene or *cistron*.

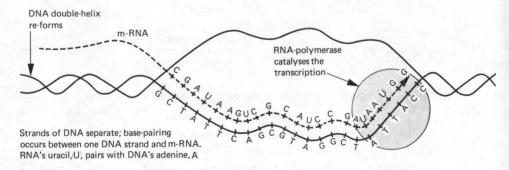

Figure 32   Transcription

The m-RNA moves to a group of ribosomes in the cytoplasm via pores in the nuclear membrane. There the message is *translated* into a sequence of amino-acids which are brought to their correct position on the m-RNA strand by molecules of t-RNA. The amino-acids link to form a polypeptide chain, the primary structure of a protein molecule. The order of bases in the DNA molecule of the gene (the genotype) ultimately produces an order of amino-acids in a protein (which determines an aspect of the phenotype). The DNA and protein molecules are said to be *colinear*: an alteration in the DNA produces an alteration in the same relative position of the protein.

Details of protein synthesis

An enzyme, *RNA-polymerase*, recognises the point on the DNA strand at which transcription should start and the point at which it should stop. These points are identified by special sequences of bases. The cistron in between is transcribed into m-RNA, the process being catalysed by the RNA-polymerase which moves along the DNA strand. Many m-RNAs may be synthesised in rapid succession along a cistron. The completed m-RNA moves away and binds to a group of ribosomes; the whole structure is called a *polyribosome*.

Free amino-acid molecules in the cytoplasm are activated by reaction with the high-energy compound *adenosine triphosphate* (*ATP*). The activation is catalysed by specific enzymes which form complexes with one kind of amino-acid and then join it to a particular molecule of transfer-RNA. The enzyme is then released. There are as many different t-RNAs as there are amino-acids. Each t-RNA molecule consists of approximately seventy to eighty nucleotides in a single strand which folds into a cloverleaf shape, leaving a triplet of bases, the *anticodon*, exposed (Figure 33).

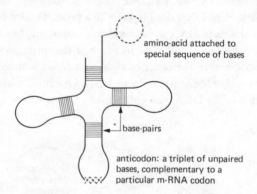

amino-acid attached to
special sequence of bases

base-pairs

anticodon: a triplet of unpaired
bases, complementary to a
particular m-RNA codon

Figure 33   A transfer-RNA molecule with amino-acid attached

The t-RNA–amino-acid complexes diffuse to the ribosomes. An m-RNA molecule has two adjacent codons available for base-pairing where it contacts the ribosome at sites 1 and 2. The ribosome moves relative to the m-RNA molecule and, each time a codon traverses site 1, a complementary anticodon belonging to a t-RNA–amino-acid complex pairs with it. The amino-acid is held in such a way that a peptide bond can form between it and the amino-acid held at the adjacent site 2 by another t-RNA. This t-RNA is then released. The two joined amino-acids (a *dipeptide*) attach to site 1 until this becomes site 2 as the ribosome moves along another codon. Gradually, the peptide chain elongates as amino-acids are brought in order, and the released t-RNAs join fresh amino-acids and repeat the process. A number of different enzymes are required to catalyse various stages in the process, but many details remain to be worked out (Figure 34).

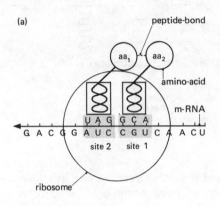

A strand of m-RNA is associated with a ribosome. At sites 1 and 2, base-pairing joins two codons of m-RNA with anti-codons of two t-RNA molecules. A peptide-bond forms between the amino-acids held by the adjacent t-RNAs.

The t-RNA at site 2 is released, leaving the dipeptide attached to the t-RNA at site 1.

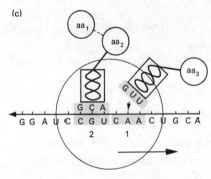

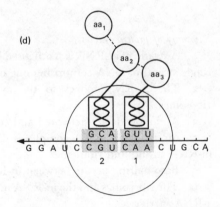

The ribosome moves relative to the m-RNA so that the t-RNA carrying the dipeptide is brought to site 2. A third t-RNA moves in to match its anticodon with the new codon at site 1.

Peptide bonding joins the newly-arrived amino-acid to the dipeptide held at site 2. The cycle repeats.

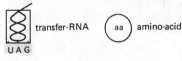

transfer-RNA   (aa) amino-acid

Figure 34   Translation

When one ribosome has moved away from the start of the message, a second attaches itself so that several polypeptide chains of different lengths will be translated from one m-RNA message at any particular time. Polypeptides may be synthesised with rapidity: the addition of one amino-acid per 0.5 seconds has been calculated. Messenger-RNAs are translated between ten and twenty times before they are broken down (see Figure 35).

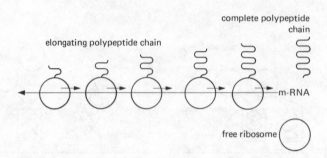

Figure 35   Polyribosome mechanism
As each ribosome moves relative to the strand, it matches a complementary t-RNA anticodon to the m-RNA codon over which it passes. Amino-acids attached to t-RNAs are brought in the order specified by the order of the m-RNA codons and linked together to form a polypeptide chain. One m-RNA strand typically associates with about five ribosomes

Summary of protein synthesis
1   A strand of m-RNA is synthesised by complementary base-pairing along a single strand of DNA comprising one cistron.
2   The m-RNA moves to the cytoplasm and associates with several ribosomes.
3   Amino-acids are activated and coupled to specific t-RNAs.
4   Amino-acid–t-RNA complexes diffuse to points of contact between an m-RNA strand and ribosomes.
5   Base-pairing occurs between m-RNA codons and t-RNA anticodons.
6   The ribosomes and the m-RNA move relative to one another as successive t-RNAs arrive.
7   Peptide bonds link adjacent amino-acids which are held in place at the m-RNA where it contacts a ribosome.
8   t-RNAs are released and recycled.
9   The finished polypeptide chain is released.

Deciphering the genetic code

By the early 1960s, attention focused on determining which triplets of bases coded for which amino-acids. The problem was enormous: a protein containing

250 amino-acids must be specified by a gene sequence of 750 bases. Whole living cells are too complicated to permit analysis, and so the technique of studying protein-synthesis in cell-free extracts of the colon bacterium *Escherichia coli* was developed. A rich culture of bacterial cells is harvested from its growth medium, centrifuged out of suspension and gently ground open. Sap containing DNA, messenger-RNA, ribosomes and enzymes is released which, when ATP is added, will incorporate amino-acids into protein.

Nirenberg and Matthaei treated such extracts with a DNA-ase enzyme which destroys DNA. They observed that protein-formation ceased after a short while, presumably when the existing m-RNA ran out. In 1961, they achieved the artificial synthesis of an m-RNA called polyuridylic acid (poly-U) which consisted of molecules containing only the base uracil. Obviously the codons borne by such a molecule must be identical: UUU, UUU. The exciting possibility was clear: if poly-U could be used to stimulate protein-formation in place of natural m-RNA, only the amino-acid for which UUU coded could be incorporated.

Nirenberg's group offered poly-U to cell-free bacterial extracts, each containing all twenty amino-acids; the extracts had previously been treated with DNA-ase. Twenty extracts were produced, each of which contained one different amino-acid labelled with radioactive carbon-14 and nineteen non-radioactive acids. In one mixture, a radioactive protein was synthesised; it contained only phenylalanine. The m-RNA sequence UUU is normally transcribed from a DNA sequence AAA, and so the DNA codon AAA stands for phenylalanine. The cracking of the code proceeded rapidly once techniques were perfected. Mixed RNA polymers containing two bases were synthesised: poly-AC, poly-AG, poly-AU, poly-CG, poly-CU, and poly-GU. Assuming that the RNA bases form a random sequence, it is possible to predict the proportions of different triplets if the ratio of the two bases is known. For instance, if poly-AG contains 60 per cent A and 40 per cent G, the probability of the triplet AAG is $0.6 \times 0.6 \times 0.4$ or 0.144. That is, 14.4 per cent of the triplets should be AAG. If 14.4 per cent of the polypeptide which is synthesised by this poly-AG consists of a particular amino-acid, it is assumed that AAG codes for it.

Laboriously, it was revealed that sixty-one of the possible sixty-four combinations of three bases specify amino-acids. This means that the same amino-acid is coded by more than one base triplet; the code is said to be *degenerate*. Usually it is only the third base of the triplet which varies. If the code were not degenerate, so that each amino-acid was specified by a unique codon, most single base mutations would be absurd and not specify any amino-acid. In a degenerate code, mutation is less likely to interfere with the amino-acid sequence, as it may result in a triplet changing to a synonym for the same acid.

Three triplets do not appear to code for any amino-acids. UAA, UGA and UAG terminate the messages. The triplet dictionary is shown in Table 5. There is considerable evidence that the code as determined by in vitro experiments functions in living organisms.

Table 5 The 'codon dictionary'

Second base

First base		U	C	A	G	Third base
U		UUU ⎫ phe UUC ⎭ UUA ⎫ leu UUG ⎭	UCU ⎫ UCC ⎪ ser UCA ⎪ UCG ⎭	UAU ⎫ tyr UAC ⎭ UAA ⎫ ct UAG ⎭	UGU ⎫ cys UGC ⎭ UGA ct UGG trp	U C A G
C		CUU ⎫ CUC ⎪ leu CUA ⎪ CUG ⎭	CCU ⎫ CCC ⎪ pro CCA ⎪ CCG ⎭	CAU ⎫ his CAC ⎭ CAA ⎫ gln CAG ⎭	CGU ⎫ CGC ⎪ arg CGA ⎪ CGG ⎭	U C A G
A		AUU ⎫ AUC ⎬ ileu AUA ⎭ AUG met	ACU ⎫ ACC ⎪ thr ACA ⎪ ACG ⎭	AAU ⎫ asn AAC ⎭ AAA ⎫ lys AAG ⎭	AGU ⎫ ser AGC ⎭ AGA ⎫ arg AGG ⎭	U C A G
G		GUU ⎫ GUC ⎪ val GUA ⎪ GUG ⎭	GCU ⎫ GCC ⎪ ala GCA ⎪ GCG ⎭	GAU ⎫ asp GAC ⎭ GAA ⎫ glu GAG ⎭	GGU ⎫ GGC ⎪ gly GGA ⎪ GGG ⎭	U C A G

The codons are triplets of m-RNA bases formed from complementary DNAs.
ct chain termination codon, phe phenylalanine, leu leucine, ileu isoleucine, val valine,
ser serine, pro proline, thr threonine, ala alanine, tyr tyrosine, his histidine,
gln glutamine, asn asparagine, lys lysine, asp aspartic acid, glu glutamic acid,
cys cysteine, trp tryptophan, arg arginine, gly glycine, met methionine

Universality of the code

Most of the research into the genetic code comes from work with microbes. There is good evidence that the code is more or less universal, however; that is, the base-sequence which codes for a particular amino-acid is the same for all species.

If transfer-RNA molecules carrying their amino-acids are extracted from the bacterium *Escherichia coli* and mixed with rabbit reticulocytes (cells that manufacture the blood protein haemoglobin), m-RNA and ribosomes, rabbit haemoglobin is formed. The codons for certain amino-acids must be the same in the bacterium and the mammal. The source of the ribosomes is irrelevant to the protein formed.

The viral way of life depends upon the virus and its host having similar codes. Experiments with the tobacco-mosaic virus, whose genetic material is RNA, show that the viral RNA directs the synthesis of enzymes and protein for the virus coat using the tobacco's amino-acids. Clearly, this could not be achieved if the codes differed. Different species may use different codons preferentially for particular amino-acids, however.

The manufacture of other kinds of RNA

The nucleolus is the site of the ribosome-manufacture, while genes for the production of transfer-RNA are scattered throughout the genome in some organisms and clustered together in others.

THE CONTROL OF GENE ACTION

Cells are able to switch their genes on and off. Bacteria are very flexible; they produce enzymes for the digestion of a new food as soon as they come across it, and stop production when that food is no longer available. Genes are now known to be controlled by molecules called *repressors* and understanding of their action comes primarily from studies of the bacterium *Escherichia coli*.

Work by F. Jacob and J. Monod in 1961 established that a series of *structural genes* synthesise the enzymes required for any several-stage biochemical pathway. These genes occur in a sequence so that information may be transcribed from them into a single m-RNA molecule which then translates into the polypeptides which make up the enzymes. Transcription is initiated by an RNA-polymerase

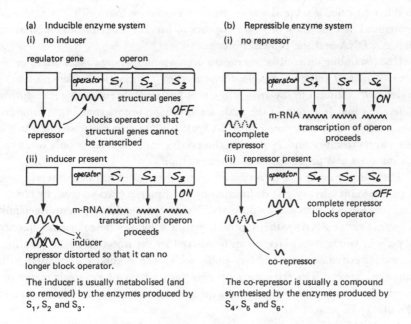

Figure 36   Operon function

molecule next to an *operator site* at the beginning of the set of genes. Jacob and Monod called the operator and its structural genes an *operon*. A separate *regulator gene* produces a repressor protein which can combine with the operator and inhibit transcription of the structural genes (Figure 36).

In *E. coli*, certain enzymes are required to digest lactose (milk-sugar). In the absence of lactose the enzymes are not required, and the genes synthesising them are switched off by the 'lac' repressor. If lactose becomes available, it acts as an *inducer*; it attaches itself to the repressor and distorts the molecule so that it can no longer combine with the operator. The transcription of the 'lac' operon can then proceed to allow synthesis of the necessary enzymes.

The control of genes can also be positive, that is, the product of a regulator gene can also keep its operon turned on until some molecular signal turns it off. Many further details await investigation, particularly concerning the control processes which operate in the cells of multicellular organisms.

Differentiation

A multicellular organism develops from a unicellular zygote to an adult with organs composed of cells with widely different physical and physiological properties. These cells derive from cell-divisions following mitotic chromosomal-divisions so that they share identical genotypes (barring mutation). They are said to *differentiate* between the zygote and the adult stage as they become more specialised for the performance of limited functions. Different genes are repressed in different cells, leaving the remainder to synthesise that cell's characteristic proteins. The repression of genes appears to involve a chemical combination between DNA and the protein histone.

Understanding differentiation means understanding how collections of genes are switched off and on as development proceeds. Hormones are key molecules which act as stimuli to the switching-on of batteries of repressed genes during differentiation. Injection of the male sex-hormone testosterone into immature rats is followed by an increase in RNA synthesis and enzyme production in the testes and prostate gland. Normally, this production would occur only when the rat matured and produced its own testosterone.

The chromosomal site of action of several hormones has been investigated using the giant salivary-gland chromosomes of dipteran (two-winged) fly larvae. Different sets of bands swell or 'puff' at different stages of development. Each puff is a site of active RNA synthesis; protein synthesis follows. The normal sequence of puffs in young insect larvae can be altered by the injection of the moulting hormone, ecdysone. Immediately puffs, which are normally associated with pupation, appear. The ecdysone level rises naturally as a larva nears metamorphosis. Presumably it acts by arousing quiescent genes needed for differentiation (Figure 37 (a)).

Another kind of giant chromosome, known as the 'lampbrush' chromosome, is found in the immature eggs of amphibians. Its name derives from thin transverse loops which appear in a regular cycle as development proceeds, giving the chromosome a bottle- or lamp brush-like appearance. These loops are thought to

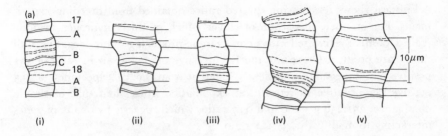

(i)   (ii)   (iii)   (iv)   (v)

Figure 37 (a) (above)   Different degrees of puffing of the band 1–17–B in *Chironimus tentans*. (i)–(iii) are from untreated animals and (iv)–(v) from larvae injected with ecdysone

Figure 37(b) (right)   Lampbrush chromosomes of a newt

(i) shows paired homologues with two chiasmata. (ii) is an enlarged view of a single loop and proposed structure.

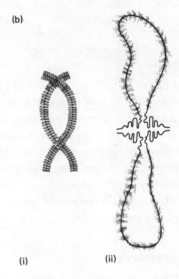

(i)   (ii)

represent despiralised sections of chromosome at which m-RNA is synthesised. They are analogous to dipteran chromosome puffs. It appears that hormones have the role of sequentially removing certain genes from the repressive action of histone (**Figure 37 (b)**).

GENE INTERACTIONS

Genes do not act in isolation; they influence each other's effects. In chapter 4, the 'position effect' was mentioned in which the extent of a gene's expression depends upon those genes which are adjacent to it. An assemblage of genes which interact to produce an integrated system is called a *gene-complex* and it may include genes whose sole function appears to be the modification of other genes' actions.

Epistasis

A common interaction in which a gene (or gene-pair) masks the expression of another non-allelic gene is called *epistasis*. The gene whose effects are masked is *hypostatic* and the suppressor *epistatic*.

Epistasis shows itself in unexpected ratios obtained from breeding experiments. For instance, if a strain of clover which produces hydrogen cyanide (called 'positive') is crossed with a non-cyanide-producing strain ('negative'), all the F_1 are positive. On selfing the F_1, investigators found an F_2 ratio of 351 positive to 256 negative, sufficiently different from the 3:1 expectancy of 455 positive:152 negative to require an explanation other than that of simple dominance. 351:256 is close to a 9:7 ratio, which suggests two pairs of genes interacting to modify a 9:3:3:1 ratio.

Hydrogen cyanide formation can be represented by:

$$\text{precursor substance} \xrightarrow[\substack{\text{enzyme 1} \\ \text{gene A}}]{} \text{cyanogenic glucoside} \xrightarrow[\substack{\text{enzyme 2} \\ \text{gene B}}]{} \text{hydrogen cyanide}$$

Enzymes 1 and 2 are produced by genes A and B, and a plant requires at least one dominant form of each of the two pairs of alleles A and B in order to synthesise appreciable amounts of hydrogen cyanide, although traces of hydrogen cyanide are produced without any dominant B allele. Gene A is epistatic to B, since B's effects are suppressed in the absence of the dominant allele of A. The number of phenotype classes is less than would be expected if the genes functioned totally independently. The positive × negative cross can be written as shown in Figure 38. Epistasis is commonly associated with the production of colour pigments in plants and animals.

Pleiotropy

While many different genes can have their effect on one phenotypic character, it is also common, perhaps universal, for a single gene to affect several characters. This situation is termed *pleiotropy*. Mendel noticed that a pea gene simultaneously affects flower colour, seed colour and the presence of a reddish colour in the leaf axils. Such pleiotropy explains why single gene mutations can often produce very drastic effects, since many processes may be upset at once. Often, the gene has one primary function in early development (pigment formation in the case cited) whose effects manifest themselves widely later on.

GENES OUTSIDE CHROMOSOMES

Chromosomal genes are inherited in equal numbers from the male and female parent; this pattern of inheritance is called Mendelian. Certain other genes are located in the cytoplasm of some cells. These are transmitted chiefly from mother to offspring, since the egg contains much more cytoplasm than the sperm. Ruth Sager has investigated the non-chromosomal inheritance of certain characters in the microscopic green alga, *Chlamydomonas*. Occasionally, non-chromosomal genes from both parents are passed to the progeny and their segregation in subsequent cell-divisions can be studied. Since this is a rare event, the

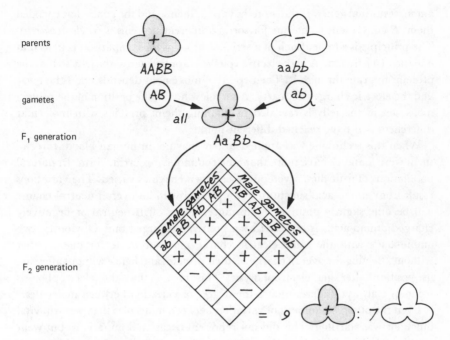

Figure 38   Epistasis

Hydrogen cyanide is formed in quantity only in those clover leaves which possess both the cyanogenic glucoside and the enzyme for its breakdown. This depends on the presence of at least one dominant allele of each of genes A and B. Gene B is hypostatic to A, since the effects of its dominant allele are masked in the absence of a dominant A

inheritance pattern is more laborious to work out than that of chromosomal genes where segregation and recombination are routine.

The cytoplasm of all plant and animal cells contains *organelles* (subcellular structures) that have their own DNA. *Mitochondria*, which synthesise high-energy compounds in both plant and animal cells, do so. *Chlamydomonas* and other green plants also contain DNA in their *chloroplasts* (organised, layered structures in the cytoplasm which contain chlorophyll pigment and synthesise carbohydrates). Chloroplast and mitochondrial DNA differ from nuclear DNA in the proportion of adenine-thymine to cytosine-guanine, which implies that it has an independent origin. Both mitochondria and chloroplasts contain the full apparatus for manufacturing their own proteins and they are possibly the sites of certain cytoplasmic genes. They divide autonomously when the nucleus is not dividing. Much remains to be discovered about the non-chromosomal genetic systems. (See also chapter 11.)

'NEUTRAL MUTATIONS'

Once it was realised that proteins are translations of the information carried in genes, geneticists had a method for discovering how much variability exists amongst normal genes. If the proteins varied, then so did the genes that encoded them. A convenient technique for sorting different proteins is *gel electrophoresis*. The principle is simple. Each of a series of proteins for comparison is placed at one end of a jelly slab. A high electric voltage is applied across the jelly so that the proteins migrate through it. Their speed of movement depends upon their size and the electric charge they carry. After a few hours, the position of the protein molecules in the jelly is revealed by staining them; proteins with molecular differences will have reached different points.

When this technique was applied to ten enzymes in human blood, three of them were found to have more than one commonly occurring form. In natural populations of fruit-flies, seven out of eighteen enzymes varied. The variations result from amino-acid substitutions; a protein with an overall neutral charge will become slightly positive, for instance, if one of its neutral or negatively charged amino-acids is replaced by a positively charged one. Obviously, two amino-acids with the same electric charge may substitute for one another without affecting the overall charge on their protein, and hence without affecting its position after gel electrophoresis. This means that the electrophoresis technique always underestimates the degree of variation between molecules.

Even the observed amount of variation between molecules that perform vital functions was startling. This did not represent occasional mutations, but well-established alternatives of molecular structure, or protein polymorphism.

By the late 1960s, the Japanese population geneticist, Motoo Kimura, was one of the first to propose that many of the molecular variants performed precisely the same function. (A convinced selectionist would maintain that each difference, however small, was the result of natural selection for subtly different functions.) Kimura argues that the mutations which become established as regular variants in a population often have no selective advantage or disadvantage: they are *neutral mutations*. He cites, as further evidence for his case, the fact that basic molecules such as haemoglobin vary in molecular composition from species to species, while performing the same task, carrying oxygen, in all of them. Two closely-related species possess haemoglobins that differ by only one or two amino-acids, while a carp and a human, whose most recent common ancestor lived 400 million years ago, have haemoglobins with only half their amino-acid sequence in common. It seems as though neutral mutations accumulate with the passage of time and cause the molecules to become progressively more different. There are limits to the amount of variation that can occur, of course, and Kimura does not suggest that all mutations are neutral. In the case of haemoglobin, it seems that only parts of the molecule can tolerate change while vital portions are rigorously maintained the same by natural selection. Kimura's conclusions have been strongly criticised by investigators who have obtained evidence for subtle differences of function between variants of certain enzymes. They maintain that notions of 'neutrality' stem from our ignorance of the complexities of the molecules' activities.

There is evidence that microbes possess a system in which mutants which would normally be selected against have neutral effects simply because they occur in 'spare copies' of vital genes. If bacteria grow on an unusual food medium, they will increase their output of any enzyme that can digest it. It has been established that the additional enzyme comes from extra copies of the gene responsible. Under normal circumstances, only one copy of the gene may be required for cell maintenance, leaving the 'spare copies' free to mutate randomly, without mutations with 'harmful' effects being selected against.

Researchers at Cambridge have grown a species of bacterium on an unsuitable food medium for thousands of generations. The bacteria which survive this treatment respond by increasing many-fold their output of any enzyme which can digest the food. The additional enzyme is produced by extra copies of a particular gene. After four years, a few bacteria began synthesising new enzymes which digested the food more efficiently, so short-circuiting the need for vast quantities of enzyme. Such evolution looks remarkably purposeful. The explanation may be that particular spare copy genes were able to mutate randomly, without suffering the consequences of selective death, because other similar genes were working normally. In time, an accumulation of random mutations might have given a functioning gene which produced a more useful enzyme. This theory is far from universally accepted, but it might explain the occurrence of far more DNA in many organisms than they appear to need for their normally-functioning genes.

QUESTIONS

1   How do we know that the genetic code for an amino-acid is a triplet of three bases?
2   On what evidence do we base the idea that the synthesis of proteins is directed by DNA?
3   In 1905, Bateson, Punnett and Saunders crossed two pure-breeding white-flowered varieties of the sweet-pea *Lathyrus odoratus* and found that all the progeny produced purple flowers. The F_1 plants were allowed to self-pollinate and, of the 651 F_2 plants which flowered, 382 had purple flowers and 269 had white flowers. Explain this as far as you can, and give the genotypes of the plants involved.
4   After irradiation of the mould *Neurospora* with ultra-violet light, several nutritional mutant forms were discovered. One mutant was found to need two supplementary amino-acids, methionine and threonine, both synthesised by wild-type *Neurospora*.
a) How could you establish that only one gene-mutation was involved?
b) Assuming that only one gene has mutated, does the mutant's need for two amino-acids necessarily invalidate the one-gene–one-enzyme hypothesis? Explain your argument.

6 Genes in Populations

Organisms rarely live in isolation, but in groups of one species called *populations* occupying a particular habitat, such as mice in a barn. Members of a population interbreed, but do not breed significantly with individuals from a different population. Natural populations of sexually reproducing organisms are genetically variable, although this is not true of asexually reproducing organisms. It is the population, not the individual, which is the unit of evolution, and the genetics of populations is therefore of vital importance.

POPULATION GENETICS

The types of breeding experiments discussed in chapter 2 are performed with parental strains which are homozygous for a given pair of alleles, say AA and aa, with genes A and a being introduced into the population in equal proportions. Only under these circumstances do the familiar 3:1 and 1:2:1 ratios emerge in the F_1. Population genetics is concerned with the frequencies of genotypes and phenotypes in successive generations of diploid, cross-fertilising, natural populations. In such cases, alleles may occur with widely different frequencies and mating does not follow the pattern prescribed for laboratory populations.

The *gene-pool* is the total number of genes available in a population at a particular time. It indicates the types of genes present and also their frequency.

The Hardy-Weinberg equilibrium

Consider a population whose members belong to genotypes AA, Aa or aa with respect to alleles A and a at one autosomal locus. The number of A alleles expressed as a percentage of the total $(A + a)$ loci in the population is called the *gene-frequency* of A. Suppose that p is the frequency of gene A and q the frequency of gene a. Since this represents all the genes at that locus in the population, $p + q = 100$ per cent, or

$$p + q = 1 \qquad \ldots (1)$$

Each egg and sperm in the gene-pool carries either A or a, so the genotypes of the zygotes which form the next generation will be:

Sperm	Egg	A	a
A		AA	Aa
a		Aa	aa

and their frequencies are revealed by using p and q:

	p	q
p	p^2	pq
q	qp	q^2

This means that the genotypes AA, Aa and aa occur in the proportions $p^2 : 2pq : q^2$. That is.

$$p^2 + 2pq + q^2 = 1 \qquad \text{(the whole population)} \qquad \ldots (2)$$

where p^2, $2pq$ and q^2 refer to the proportion of dominant homozygotes, heterozygotes and recessive homozygotes respectively.

Notice that equation 2 is a binomial expansion of $(p+q)^2 = 1$. The situation from which Mendel's first law is derived is a special case of this binomial square law, in which $p = q = \frac{1}{2}$.

The basic relationship between gene-frequency and zygote-frequency was observed independently by G. H. Hardy, a British mathematician, and W. Weinberg, a German physician, in 1908. These men noted that gene-frequencies, p and q, remain theoretically constant from generation to generation. There is no innate tendency for a dominant gene to become more common at the expense of its recessive allele, or vice versa. This principle is the basis of population genetics.

An example illustrates the maintenance of the Hardy-Weinberg equilibrium. In human populations, carriers of the dominant gene, R, can roll their tongues up sideways, whilst homozygous rr people cannot. Suppose that, in a large, isolated population, non-rollers outnumbered homozygous rollers in a ratio 4rr : IRR, with no heterozygotes present. The gene-frequencies in this gene-pool will be eighty per cent r and twenty per cent R or 0.8r : 0.2R.

Assuming that individuals choose their mates without regard for their tongue-rolling ability, the genotype of the next generation will be made up as follows:

Eggs	Sperm	0.8r	0.2R
0.8r		0.64rr	0.16Rr
0.2R		0.16rR	0.04RR

That is, sixty-four per cent of individuals will be homozygous non-rollers, thirty-two per cent heterozygous rollers and four per cent homozygous rollers. The relative proportions of gametes from this generation will be:

r 0.64 (from rr) + 0.16 (from Rr and rR) = 0.80
R 0.16 (from Rr and rR) + 0.04 (from RR) = 0.20

In other words, the gene-frequency of $0.8r : 0.2R$ recurs, as does the phenotype ratio of 64 non-rollers : 36 rollers. Notice that, in this hypothetical example, the recessive character is more common than the dominant one because more non-rollers than rollers initiated the isolated population. 'Dominance' does not always imply 'predominance'. In real human populations the ability to roll the tongue is generally more common than the inability to do so.

Where complete dominance occurs, heterozygotes and dominant homozygotes are phenotypically indistinguishable. It is sometimes important to be able to calculate the proportions of the two different genotypes within the dominant phenotype. Albinism, for example, with its lack of pigment in eyes, skin and hair and other disabilities, occurs in about one person in 20000. Pedigree studies show that the gene for albinism is recessive, therefore $q^2 = 1/20000$ or 0.00005. The frequency of the recessive allele, q, is $\sqrt{(0.00005)}$ or 0.0071, and that of the dominant allele, p, is $1 - q$, or 0.9929.

$$p^2 = 0.9929 \times 0.9929 = 0.9859 \qquad \text{or } 98.59\%$$

$$2pq = 2 \times 0.9929 \times 0.0071 = 0.01410 \qquad \text{or} \qquad 1.410\%$$

This means that approximately one person in seventy-two is a heterozygous carrier of the albinism gene – a surprisingly high proportion, considering the rarity of the recessive allele.

If natural populations behaved exactly according to the Hardy-Weinberg principle, evolution (*change* in the genetic composition of populations) would be impossible. The dynamics of natural populations include many factors which disturb the genetic equilibrium. When any of these factors is present, the Hardy-Weinberg 'law' no longer applies exactly.

Factors influencing gene-frequency

1 Mutation
If gene A recurrently mutates to a, the gene-frequencies p and q must change. Assuming a to be as fit to survive as A, A would eventually disappear from the population. Mutation rates of most genes are extremely low, however, so the chance of this happening, particularly in organisms with long generation times, is remote. Moreover, reverse mutations $(a \rightarrow A)$ usually also occur, but not necessarily at the same rate, so that an equilibrium is established.

2 Migration
Gene-frequencies can be altered if members emigrate or if foreign individuals with different genotypes immigrate into the population.

3 *Natural selection*

Darwin's theory of natural selection depends upon the fact that some combinations of alleles are more conducive to survival and reproduction than others. The possessors of different genotypes do not transmit their genes to the next generation with equal frequency as the Hardy-Weinberg law requires. Those individuals which survive and become parents form a non-random sample of the population.

4 *Genetic drift*

The number of individuals carrying a particular allele will vary somewhat from generation to generation so that gene-frequencies fluctuate about a mean. This is because mating will not be entirely random and the gametes which unite to form the progeny are a mere sample from the entire parental gene-pool. Sampling error therefore produces a random shift of gene-frequencies or *genetic drift*. In a very large population, such drift will have negligible effects but, in a small population, the chance of non-representative progeny becomes greater.

Summary of the Hardy-Weinberg law

The Hardy-Weinberg law describes the equilibrium which exists between alleles in a theoretical large population in which random mating occurs in the absence of mutation, migration, selection and genetic drift. It demonstrates the important point that there is no inherent mechanism in Mendelian inheritance for altering gene-frequencies. The factors which change the Hardy-Weinberg equilibrium are those which alter gene-frequencies and hence lead to evolution.

WAYS OF BREEDING

The Hardy-Weinberg law applies only to cross-fertilising species. There are other types of breeding systems which result in different amounts of population variability and hence different responses to the influence of natural selection. These systems will be discussed, starting with those which produce least variability.

Asexual reproduction

The vast majority of multicellular organisms are diploid; every gene has an allelic partner. Should one member of a pair of alleles mutate, there will usually be no obvious phenotypic change in that organism, since most mutant genes have effects which are recessive to those of the 'normal' gene. *Asexual reproduction* depends upon mitotic division in which genes are duplicated exactly to form progeny with genotypes identical to that of their parent and to those of one another. It follows that diploidy confers genetic stability in asexual reproduction, since it prevents recessive mutations from showing phenotypic effects. Polyploids, with several sets of each gene, are even more stable. In haploid organisms, such as bacteria, a mutation in any of the single set of genes will find phenotypic expression immediately. For them, asexually doubling and dividing can allow genetic flexibility because they multiply rapidly and mutants are quickly propagated. In a diploid organism, however, a mutation must have dominant effects in order for these effects to show, or else (and this is extremely unlikely) identical mutations must occur simultaneously in both alleles.

Multicellular organisms which rely exclusively on asexual reproduction are genetically rigid. They may be very successful in a particular environment, since they can spread rapidly when conditions are favourable without the need to obtain a partner for mating. Constant high-yielding strains of plants such as potatoes and fruit-trees are maintained by horticulturalists using techniques of asexual propagation. If the plants were allowed to reproduce sexually, the offspring would not be of uniformly good quality. Asexual reproduction has the advantage of preserving valuable gene combinations. Should environmental conditions change, however, there is no reservoir of variety in a genotypically identical population to provide differently adapted forms. In natural populations it is common for asexual and sexual methods of reproduction to alternate.

Sexual reproduction

Sex provides the variability that is needed for diploid organisms to evolve to suit new conditions. The two fundamental processes in sexual reproduction are meiosis and fertilisation. In meiosis, the independent assortment of chromosomes and crossing-over reshuffle the genes into different arrangements, while fertilisation brings new combinations together. Mutants that arose in separate individuals can therefore be assembled in one organism. This genetic recombination is vitally important in providing an immediate source of variation on which natural selection can act. Ten mutations in an asexually reproducing population could result in only ten novelties, even if the effects were phenotypically expressed. Ten mutations in a sexually-reproducing diploid population could give $2^{10} = 1024$ novelties. To quote the famous geneticist, Dobzhansky, : 'sex arose in organic evolution as a master adaptation which makes all other evolutionary adaptations more readily accessible'. Indeed, the recombinational ability which is fundamental to sexual reproduction is a basic property of DNA itself.

Inbreeding

Flowering plants are usually *hermaphrodite* (having *stamens*, which produce the male pollen, and *carpels*, which contain the female eggs, in the same flower) or *monoecious* (having separate male and female flowers on the same plant). In some plants, the closeness of male and female gametes leads to self-fertilisation, a form of reproduction which allows the recombination of genes belonging to that one plant only. The common pea self-pollinates and the result is a plant whose offspring's uniformity is similar to that of the offspring of asexually reproducing plants. This is the reason why Mendel could use his parental 'pure lines' with the knowledge that they would always 'breed true'. Self-fertilisation is the extreme

form of ~~*~~*reeding* or breeding between related individuals. It is so effective in
prod...   ...formity that, if a population of peas were artificially cross-
... ...heterozygous (Aa) seeds and afterwards their descendants self-
...   ...would...take only four more generations before ninety-seven per cent
of all the offspring were homozygous (AA or aa). Try to verify this for yourself,
remembering that, whilst homozygotes breed true, self-fertilising heterozygotes
produce 1 dominant homozygote:2 heterozygote:1 recessive homozygote offspring. Inbreeding, then, produces a population whose individuals have very
similar genotypes, highly suited to the environment. They may survive
extremely well, yet lack the ability to evolve rapidly to suit new conditions.

Outbreeding

Outbreeding is breeding between unrelated individuals. The first generation
after outbreeding two different strains, each of which was previously inbred, is
frequently unusually vigorous. This 'hybrid vigour' is exploited in agriculture
when F_1 seeds of plants such as sweetcorn and tomatoes are commonly used to
give productive crops. The reason for the exceptional vigour of the F_1 plants has
been explained as follows. The two parental inbred lines have accumulated
different alleles promoting good growth. These may include genes for extensive
root systems, much chlorophyll and earlier germination. In the inbred lines,
these genes are found in the homozygous condition and each line has a different
collection. On crossing, the F_1 will be heterozygous for each pair of alleles and so
gain the benefit from all of them, since most advantageous alleles express
dominance. Inbreeding the F_1 generation does not, of course, lead to a uniform
F_2.

Outbreeding maintains more variety in the gene-pool than does inbreeding
and favours heterozygosity. It promotes evolutionary flexibility at the expense of
producing progeny not equally well-suited to the prevailing conditions.

Outbreeding mechanisms

Mechanisms which ensure that outbreeding occurs have developed in many
organisms. Self-fertilisation can be prevented by having two sexes, as most
animal species do. Many flowering plants which are hermaphrodite cannot self-
fertilise. Pollen grains may ripen before or after the stigmas are receptive,
keeping pollen and eggs apart in time. Often the plant's pollen is incapable of
growing down its own style to effect fertilisation: it is said to be *incompatible*. Such
incompatibility is genetically controlled by a series of multiple alleles. A diploid
plant carries any two alleles of a series $S_1, S_2, S_3, \ldots$ and is usually heterozygous
at the S locus. An S_1S_2 plant will produce S_1- and S_2-type haploid pollen and
neither type will germinate on its own stigma. Only pollen with a different
genotype, say S_3, which must come from a different, probably unrelated, plant
can effect fertilisation (Figure 39).

Many flower structures have evolved which ensure that pollen produced by
the anthers cannot reach the same flower's stigma. Primroses, for example, occur
in either of two forms, called 'pin-eyed' and 'thrum-eyed', whose anthers and

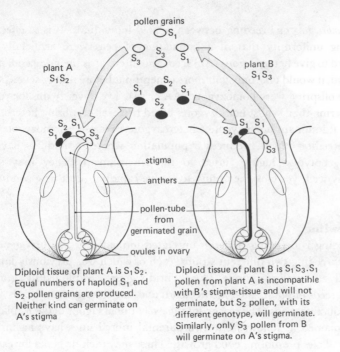

pollen grains

plant A
S_1S_2

plant B
S_1S_3

stigma

anthers

pollen-tube
from
germinated grain

ovules in ovary

Diploid tissue of plant A is S_1S_2.
Equal numbers of haploid S_1 and
S_2 pollen grains are produced.
Neither kind can germinate on
A's stigma

Diploid tissue of plant B is S_1S_3. S_1
pollen from plant A is incompatible
with B's stigma-tissue and will not
germinate, but S_2 pollen, with its
different genotype, will germinate.
Similarly, only S_3 pollen from B
will germinate on A's stigma.

Figure 39 Sections through stylised flowers which exhibit self-incompatibility
mechanisms

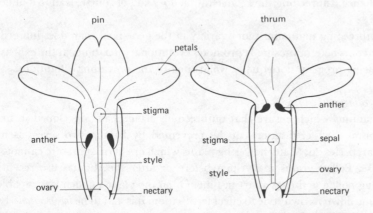

pin

thrum

petals

stigma

anther

anther

stigma

sepal

style

style

ovary

ovary

nectary

nectary

Figure 40 Pin- and thrum-eyed primroses (half-flowers) (natural size)
Pollen from a primrose rubs on to the head and tongue of an insect, such as a bee,
when it reaches into the flower for nectar. Pin-pollen brushes on to the insect in a
position suitable for rubbing on to the stigma of any thrum-primrose that the insect
subsequently visits. Similarly, pollen collected from a thrum-flower is positioned
where it can be rubbed on to a pin-stigma. Primroses are further safeguarded
against accidental self-pollination since pin-pollen will not readily germinate on a
pin-stigma, nor thrum-pollen on thrum-stigmas

stigmas are positioned differently and facilitate cross-pollination by visiting insects (Figure 40).

Breeding programmes

In modern agriculture, crops and livestock have been bred for high yield and uniformity. Horticulture depends largely upon self-fertilised plants, such as wheat, to give uniform crops. Animal husbandry relies on controlled inbreeding, however, which has different genetic consequences. Self-fertilising crops long ago reached almost complete homozygosity, so that any harmful effects from recessive mutant alleles have manifest themselves in the phenotype already and have been eliminated by natural selection. Artificial inbreeding of animals, which are normally outbreeding, brings together mutant recessive alleles for the first time. (Close relatives are much more likely than unrelated animals to have identical mutant genes derived from a common ancestor.) While the mutant's effects would be masked by the dominant allele in the heterozygous condition, in the 'double-dose' homozygous condition it may prove harmful. During an inbreeding programme, it is necessary to cull from the herd any weakly or deformed individuals to ensure that those with an obviously unfit genome do not breed.

Depletion of genetic resources

Similar varieties of domestic plants and animals are cultivated in a large proportion of the world, these relatively new introductions replacing the huge variety of wild types that had evolved in their natural habitats. In the wild state, a rich variety of mutant genes in the gene-pool is an asset in an outbreeding population, since some of their effects may prove useful to the organism's survival under changed conditions. The price paid for this asset is an occasional unfit gene combination, which is eliminated by natural selection. This elimination happens much more rapidly in a programme of inbreeding and culling and results in a serious depletion of variety in the gene-pool.

Now the dangers of depleting the genetic resources of both wild and domestic species are becoming more widely appreciated. Artificially 'improved' varieties introduced in crop *monoculture* (the cultivation of only one crop) and intensive animal production, result in populations of a few genotypes only. These are vulnerable to pests and pathogens (which themselves tend to evolve rapidly) to which they can offer little resistance. Changed environmental conditions, too, can result in decimation of populations which have only a narrow spectrum of variability.

These disadvantages of the inbred domestic breeds are particularly apparent when they are introduced into parts of the world quite different from those for which they were artificially selected. Under unaccustomed climatic conditions, faced with different pests, predators and competitors, the qualities for which the introduced varieties were selected often have little opportunity to show themselves.

Alongside the modern tendency to concentrate agriculture on fewer and fewer domestic types, is the tendency to deplete the numbers of wild species. Destruction of natural habitats is the most common reason for the decline and extinction of numerous species of plants and animals. Moreover, the obvious vertebrate animals and vascular plants which are disappearing represent only the tip of the iceberg. Small organisms which play a vital rôle in the maintenance of soil fertility, for example, may be destroyed before they are even recognised. Quite apart from ethical and aesthetic considerations, there are genetic reasons why such destruction is damaging to human chances of survival. The wild relatives of cultivated plants and animals form important reservoirs of genetic material with which domestic varieties may be crossed and so improved to suit local conditions. Moreover, the wild species are evolving continuously, meeting the changing demands of their habitats, so that it is vital to conserve the species in their natural habitats.

Domestic varieties, too, were at one time more diverse than they are today. Between eighty and ninety breeds of cattle, sheep and pigs with unique genetic combinations have become extinct in Britain to our knowledge. Many of them were capable of withstanding harsher conditions than can the modern high-yielding breeds.

Conservation

The Food and Agriculture Organization (FAO) is active in the field of conservation of genetic resources through the work of its Crop Ecology and Genetic Resources Unit. By means of symposia and advisory committees, the extent of the problem and some possible solutions are becoming known. Expeditions are mounted to various parts of the world to collect seeds, for instance of wild cereals and legumes. In many cases, particularly where destruction of the habitat cannot be avoided, priority is given to establishing artificial plantations of the crop.

In 1973, the FAO launched a pilot project on Conservation of Animal Genetic Resources, in co-operation with the United Nations Environment Programme (UNEP). The resulting surveys indicated the value of local breeds. In Botswana, for instance, the local Tswana cattle had better growth-rates and fertility, when kept under slightly better conditions than normal, than the introduced cattle which were bred in South Africa. Cross-breeds, with the advantage of hybrid vigour, might be an even better proposition. The West African Shorthorn cattle are particularly important in their locality, since they have a genetic tolerance to trypanosomiasis, a disease caused by a protozoan and spread by tsetse flies. Introduced cattle are exterminated by the disease unless costly precautions are taken.

QUESTIONS

1 Of 100 students, 64 could roll their tongue lengthwise. Tongue-rolling is governed by genes at a single locus.
   (a) What is the gene-frequency of the recessive allele in the population?

(b) What is the gene-frequency of the dominant allele?

(c) How many of the group are heterozygous for this gene?

2 4500 babies were born during one year in a group of hospitals and 9 of them died of a pancreatic disease. This disease results from the action of a single recessive gene.

(a) Work out the gene-frequency of the mutant allele.

(b) What proportion of the population are (i) homozygous and (ii) heterozygous for this gene?

3 A serious human metabolic disorder is heterozygous in approximately 0.04 per cent of the population. Mary, who is planning to marry, had a sister who died of the disease. Work out her chances of bearing affected children.

4 Discuss the methods of reproduction which produce:

(a) the most variable offspring

(b) the least variable offspring.

To what circumstances are each of them best suited?

5 Explain what is meant by the expression 'the depletion of genetic resources of a species'.

7 Genetic changes in populations

Evolution may be defined as genetically inherited cumulative change in the characteristics of interbreeding populations of organisms during the course of successive generations. Those factors which disturb the genetic equilibrium of a population – mutation, immigration and emigration, natural selection and genetic drift – are those which may result in evolution. Mutation and immigration increase the genetic variety of a gene-pool while emigration, natural selection and drift tend to decrease that variety. All except natural selection act randomly. Natural selection operates by the differential survival and reproduction of those organisms whose phenotypes are the best adapted to their conditions of life. Since the phenotype depends in part on the genotype, natural selection indirectly selects certain genotypes. It is the only force capable of increasing the overall adaptation of a population to its way of life.

Johannsen's experiment

Phenotypic variation between members of a population can result from environmental or genetic differences or from a combination of the two. Early this century, W. Johannsen demonstrated that selection can induce evolutionary changes in a population only if some of the variation is of genetic origin.

Johannsen used separate pure lines of beans which, like Mendel's peas, were homozygous for all genes after many generations of self-pollination. In each line, Johannsen selected the smallest and the largest beans to plant for the next generation. After many generations of this artificial selection, the average weight of beans from the 'heavy' parents was the same as that from the 'light' parents. The individuals all had identical genotypes and so phenotypic differences were entirely of environmental origin. If the same experiment is performed on the progeny from cross-pollinations from different pure-lines of beans, selecting the heaviest beans gives a race with a heavier mean than that of the original parent generation, while selecting the lighter beans gives a lighter race (Figure 41). *Selection is effective only when there is a range of genetic variability on which to act.*

The evolution of heavy strains of peas is an example of *directional selection* in which the extreme expression of the phenotypic character concerned is most favoured. (This happens to concern selection brought about 'artificially', i.e. by human intervention, but directional selection occurs commonly in nature as well.) Paradoxically, natural selection in wild populations may act to favour the mean phenotype; this is *stabilising selection* and it tends to resist evolutionary change. Under other circumstances, natural selection favours extremes in

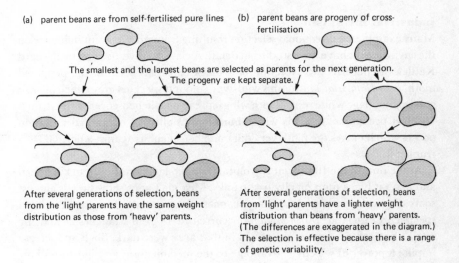

(a) parent beans are from self-fertilised pure lines

(b) parent beans are progeny of cross-fertilisation

The smallest and the largest beans are selected as parents for the next generation. The progeny are kept separate.

After several generations of selection, beans from the 'light' parents have the same weight distribution as those from 'heavy' parents.

After several generations of selection, beans from 'light' parents have a lighter weight distribution than beans from 'heavy' parents. (The differences are exaggerated in the diagram.) The selection is effective because there is a range of genetic variability.

Figure 41   Selection for weight in beans

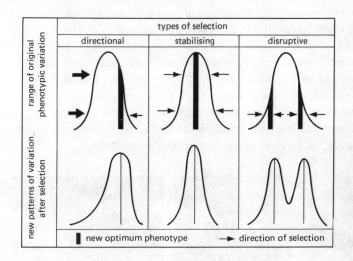

Figure 42   Types of selection
The effect of selection may be to alter the pattern of variation in succeeding generations

preference to the mean so that different forms exist together; this is termed *disruptive selection* (**Figure** 42).

Directional selection

For directional selection to occur in a population, the environment must change so that the population's gene-pool is no longer ideal under the new conditions and a new optimum gene-pool is selected.

Industrial melanism

Many examples of directional selection resulting from man's recent influence on the environment have been studied in detail. A classical investigation by Bernard Kettlewell concerns the evolution of a black or *melanic* form of the peppered moth, *Biston betularia*. The moth's wild-type colour is speckled grey which offers it good camouflage while resting with wings spread on a lichen-covered tree-trunk. Its chief predators are birds which hunt it by sight. An occasional mutation occurs which makes the moth very dark and conspicuous against a normal tree-trunk.

At the time of the Industrial Revolution, smoke-laden air began to kill lichen and deposit black films on city tree-trunks. Melanic moths show up less against sooty tree-trunks than do wild-type ones (Figure 43). The first recorded specimen of melanic *Biston betularia* occurred in 1848 in Manchester and, by 1895, about 98 per cent of the moths in that area were dark. Such an increase implies a powerful selective advantage to the melanic form and illustrates how rapidly gene-frequencies in a population can change under such circumstances. Kettlewell tested the hypothesis that wild-type forms survived bird-predation better in a clean atmosphere while the melanic form survived better in polluted areas. He released equal numbers of wild-type and melanic moths in a Dorset wood where atmospheric pollution was minimal and the trees were lichen-covered. Observations from hides revealed that birds captured 164 melanic moths but only twenty-six wild-type ones, presumably because of the superior camouflage of the wild-type ones. An identical experiment carried out in Birmingham, where tree-trunks were blackened with soot, produced opposite results. Here, redstarts alone ate forty-three wild-type moths and only fifteen melanic ones. Predatory birds were acting as agents of natural selection, favouring the survival of the best-camouflaged moths.

Figure 43   The two forms of *Biston betularia* (the peppered moth) on normal and sooty tree trunks

The change in the atmosphere from clean to polluted is associated with the spread of the melanic gene through the population as the 'selection-pressures' alter in its favour. Since genes generally have multiple effects, it is possible that an invisible physiological or biochemical effect of the mutant gene is at least as important as the colour change. Indeed, it has been demonstrated that the gene responsible for melanism in the adult confers on the larva an increased capacity to survive on limited food. Melanic forms survive better, except when the larval advantage is over-ridden by the adult camouflage disadvantage. This phenomenon of 'industrial melanism' is widespread among insects.

Stabilising selection

The young of a species tend to be much more variable than the adults, and it is usually the extreme variants which fail to mature. Stressful conditions tend to destroy all but the mean of those which survive to adulthood, as a report by H. C. Bumpus in 1898 demonstrated. Bumpus collected a group of stunned sparrows, brought down by a snow-storm in America. He measured characters such as wing-spread and weight and discovered that the measurements of those which survived tended to be closer to the mean than those of the birds which died. It seemed that the variation that could be tolerated while conditions were less taxing was not permissible when selection was intense. The differential mortality of the sparrows is an example of *stabilising selection* (assuming that their variability had some genetic basis). This is probably the most common way in which natural selection operates. Organisms which have lived for millions of years in very stable environments closely resemble their ancient relatives. The oyster and the coelacanth fish are examples of creatures which have retained an almost constant form, presumably by the operation of stabilising selection.

Disruptive selection

If selection pressures favour the extremes of variation so that intermediates are selected against, the result can be the evolution of two sub-populations, or *races*. Should the trend continue, the races may become so different that they no longer interbreed and two separate species are formed. Such species-formation, or *speciation*, will be considered later in the chapter.

Polymorphism

Another effect of disruptive selection is to maintain within one interbreeding population a number of distinct types of individual or *morphs*. The proportion of the rarest morph is greater than can be maintained by recurrent mutation alone; balanced selective forces operate, some favouring one morph and others the alternatives. Such a stable situation is called a *balanced* or *stable polymorphism*, to distinguish it from *transient polymorphism* such as the gradual increase of the melanic form of *Biston betularia* at the expense of the wild-type. The two forms of moth are present together in appreciable numbers only temporarily while one gene is in the process of replacing the other in the gene-pool.

Frequently, balanced polymorphism results from the advantage of hetero-zygotes over homozygotes. Heterozygotes have the benefits of genes whose advantageous effects have, through evolution, become dominant. These same genes have other disadvantageous effects which, being recessive, do not manifest themselves in the heterozygote. To take an extreme example, suppose an organism heterozygous for a pair of alleles, Aa, breeds effectively, while the homozygotes AA and aa are sterile. Each generation, the Aa individuals reproduce to produce the genotypes AA : Aa : aa in the proportion 1 : 2 : 1, thus maintaining the polymorphism. There is no way in which the fertile Aa individuals can reproduce their own morph only.

Sickle-cell anaemia

A well-known example of a polymorphism maintained by heterozygote advantage comes from a study of the blood pigment haemoglobin in humans. Most adults' haemoglobin is a type called A, controlled by a gene Hb^A. In Africa, a mutant allele, Hb^S, occurs which controls the production of haemoglobin-S. Heterozygotes Hb^AHb^S produce some haemoglobin-A and some haemoglobin-S. Such blood appears normal in the body but, if abnormally low oxygen pressures occur, the red cells distort or 'sickle' (Figure 44). In a homozygous person, Hb^SHb^S, all the haemoglobin is type-S and this causes sickling of the red cells even at oxygen pressures found normally in the body. The collapsed cells block small blood-vessels and are destroyed by the body, resulting in severe and often fatal anaemia.

The Hb^S gene attains frequencies of twenty per cent in parts of East Africa, despite drastic selection pressures which eliminate eighty per cent of Hb^SHb^S children before they are old enough to reproduce. The reason is that heterozygotes, Hb^AHb^S, survive the effects of malaria better than do normal Hb^AHb^A homozygotes and leave proportionately more offspring. The hetero-zygotes enjoy an advantage over both homozygotes, one of which tends to die in infancy from sickle-cell anaemia and the other from malaria. A balanced

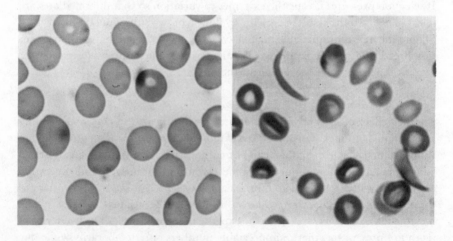

Figure 44   Normal and sickle-cell red blood cells

polymorphism for these genes exists in areas where malaria is endemic, so that as many genes are lost because of the disadvantage of the homozygotes Hb^SHb^S as are gained because of the heterozygote advantage.

Selection for outbreeding

The mechanisms, discussed in chapter 6, which maintain outbreeding by ensuring that mating is *not* random, also automatically ensure polymorphism. The many differences which distinguish pin-primroses from thrum-primroses are controlled by a series of closely-linked genes (rarely separated by crossing-over) called a *super-gene*. The super-gene S, which determines the thrum characters, behaves as a single gene whose effects are dominant to s, which determines pin characters. Thrums cannot fertilise thrums and these are always heterozygous, while pins are recessive homozygotes. The usual thrum-pin cross, Ss × ss, results in approximately equal numbers of pin and thrum offspring, generation after generation.

Colour variation in Cepaea

The British land snails *Cepaea nemoralis* (brown-lipped) and *C. hortensis* (white-lipped) exhibit polymorphisms of shell-colour and banding patterns that have been thoroughly investigated, particularly by A. J. Cain, P. M. Sheppard and J. D. Currey. In *C. nemoralis*, a series of multiple alleles controls shell colour. The dominance order is brown, pink, yellow, brown being 'top-dominant'. Plain, unbanded shells are controlled by genes dominant to those which control dark bands. Up to five bands may be present and sometimes bands are fused together. Some of the commonest patterns are shown in Figure 45.

Cain and Sheppard compared the frequencies of different phenotypes in different habitats and demonstrated a tendency for those types which blend well with their background to be most common. Woods carpeted with decaying leaves housed high proportions of plain pink and brown shells whereas, in grass and hedgerows, yellow banded snails were most numerous. The snails are preyed

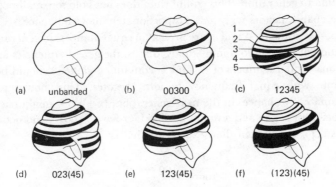

Figure 45   Polymorphic forms of the brown-lipped snail, *Cepaea nemoralis* (natural size)
Bands are numbered from the top of the largest whorl downwards. A missing band is marked 0. Two bands fused together are bracketed

on by birds, particularly song-thrushes, which smash their victims on suitable stones and eat the flesh. The broken shells remain at the thrushes' 'anvils'. Collection of the pieces revealed that they were not a random sample of the snails living in the area, but contained a disproportionately high number of shells which were least well camouflaged to a human observer. It appears that visual predation is an important selective pressure governing the snails' survival.

Polymorphism implies balanced selective pressures; if predation were the only controlling factor we would expect uniformly well-camouflaged snails in each habitat. Cain and Currey have shown that snail populations in exposed and taxing environments may not contain many well-camouflaged individuals. They found that yellow shells with few bands are common under these conditions and have suggested that the genes concerned also confer some physiological advantage. It is probable that heterozygotes also have some superior fertility or viability, thus ensuring that populations remain polymorphic. Other selection pressures adjust the proportions of types in different ways under different circumstances. Occasional drastic changes in population composition over short distances of apparently fairly uniform habitat have been recorded. This suggests that subtle changes of micro-habitat may have selective influence on some phenotypic expression of genes which also happen to control colour and pattern. Much remains to be elucidated.

GENETIC DRIFT

Natural selection is the agent which is responsible for most evolutionary change by reducing and channelling genetic variation. Another factor which can cause one mutant to become common in a population, at the expense of its allele, is the random fluctuation in gene-frequency called genetic drift. In chapter 6, it was explained that, in large, randomly interbreeding populations in which there is no mutation or selection, there will be no change in gene-frequency from generation to generation. This equilibrium does not hold for small populations because chance factors will exert proportionately more influence.

Suppose two alleles, A and a, are present in equal numbers in generation n in a population of 1000 diploid individuals, that is, the gene-frequencies are 1000A and 1000a. Equal numbers of A- and a- containing gametes should be formed and unite with equal frequency to form zygotes. This, however, is only approximate. By chance, in the next generation, $(n+1)$, A might occupy four more loci than before and a might occupy four fewer: 1004A and 996a. A now appears 0.4% more often than expected. That is:

$$\frac{4}{1000} \times 100 = 0.4\%$$

In a small population of ten individuals, a similar divergence would give fourteen A and six a alleles in generation $(n+1)$ – an excess of 40 per cent A. Supposing that, in both cases, A and a have the same selective advantage (that is, neither improves its possessor's chances of survival or reproduction more than does the other), the gametes that go to form the subsequent generations will be in proportion to the frequency of A and a in the gene-pool. Should chance increase the disparity between the numbers of A and a alleles, it is possible that a might disappear within a few generations so that all the progeny are homozygous AA. Gene A is then said to be *fixed*. No further change in this gene-frequency is possible unless the lost allele is replaced by mutation or immigration. The genetic variation of the population has been reduced. Drift, like inbreeding, tends towards genetic uniformity. In the large population, it is highly unlikely that chance variation would favour one allele for sufficient generations to eliminate the other. It is much more probable that the allele frequencies would fluctuate about the mean.

This principle has important consequences for populations which fluctuate in size as well as those which remain permanently small. An allele which is lost owing to genetic drift in a reduced population remains lost when the population increases, unless it is replaced by mutation or immigration.

The founder principle

When a few emigrants leave a parent population and found a new colony, they will not carry genes representative of the whole original gene-pool. Their range of genetic variation will be limited and largely randomly chosen. Different founding colonies will possess different genetic compositions which will influence the path that evolution by natural selection subsequently takes. This particular case of genetic drift is called the *founder principle*. It explains the divergence between small, isolated populations of a species which have been established by a very few colonists. Darwin's finches provide a very good example.

GENE-FLOW AND EVOLUTION

The movement of alleles within and between populations is termed *gene-flow*. The movement may be extensive in a highly mobile, outbreeding species or almost non-existent in a sessile, inbreeding one. The ideal condition for rapid evolution seems to be that in which an abundant species is sub-divided into units with only limited gene-flow between them. Both drift and selection may then operate to produce different gene-pools in each partly-isolated population.

W. H. Dowdeswell and E. B. Ford are pioneers in the science called *ecological genetics* which combines genetics and field-work. They studied populations of the meadow-brown butterfly, *Maniola jurtina*, on the Isles of Scilly and made comparisons between the variability of butterflies collected from large and from small islands. The hind wings have up to five black spots on the underside and the distribution of spot-numbers varies between the sexes and between different populations (Figure 46). Dowdeswell and Ford chose to study this character because it is easy to assess, and the differences being quantitative rather than qualitative suggests that the character is under polygenic control and therefore quick to respond to the effects of natural selection. There is no evidence that the

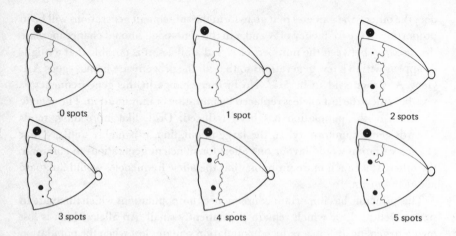

Figure 46   The meadow-brown butterfly, *Maniola jurtina* (natural size)
The undersides of the hind-wings are revealed to show the various numbers of spots near the edge

spot number itself is of significance to the butterfly, but it appears that the controlling genes have other effects which are important. Butterflies from the three largest islands (270 ha = 680 acres) have a uniform spot-distribution whose small variations are not statistically significant. Populations from islands of less than 16 ha (40 acres) show great diversity of spot-distribution (Figure 47). Ford and Dowdeswell discount the role of drift in establishing the differences between the small populations. Their explanation is that small, isolated populations become adapted by the action of natural selection to the distinct ecological features of their habitat. The small Scilly Isles differ notably from one another ecologically. Butterflies inhabiting larger islands must adjust to the wide range of different habitats present: these are similar on all islands. These butterflies therefore develop an optimum average spot-distribution. This reasoning is sound only if the butterflies range across the whole of the large islands and form a freely interbreeding population on each one. Whatever the correct explanation, it is undeniable that the smaller populations differ more from one another than do the larger ones, and that the differences are genetically controlled. Since evolution depends on the accumulation of different genes, it clearly happens faster in the smaller isolated populations. This principle does not extend to extremely small populations, since the depletion of the gene-pool by inbreeding and drift then becomes excessive.

Observations on the numbers of individuals present in a particular area do not tell everything about the effective size of a breeding population. Social constraints may operate between animals and so limit gene-exchange. The behaviour which establishes a hierarchy among male mice, for instance, ensures that the dominant male of each social unit sires about ninety per cent of the litters. Inbreeding and drift must operate to an extent which could not be judged merely by counting mice.

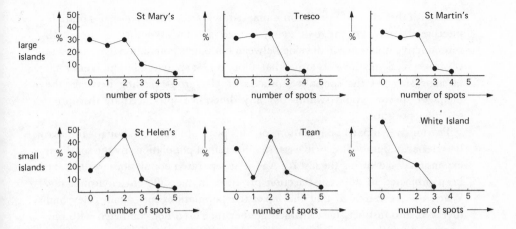

Figure 47   Distribution of spot-numbers of the meadow-brown butterfly, *Maniola jurtina*, on the Isles of Scilly

Spot number distributions on the large islands resemble one another much more closely than do those on the small islands. This implies that populations on each of the small islands are evolving in a slightly different way from those of the other small islands. Large-island populations evolve so as to resemble one another, possibly because each island has a range of similar habitats to which the butterflies must adapt

SPECIATION

Species

The meaning of the word 'species' has changed since the earliest days of biological classification, or *taxonomy*. Originally a species was merely a rank in the scheme of classification which contained animals or plants which closely resembled one another. Since the members of such 'species' were frequently identified by examination of corpses, the decision concerning which to include was often arbitrary. Modern taxonomy defines the species as a group of organisms which (at least potentially) forms an interbreeding collection of populations unable to breed freely with other species. There are drawbacks to this definition, since it cannot include those organisms which are extinct or those which do not reproduce sexually and so their species-affinities must be assessed differently. The definition has the advantage of being biologically meaningful for sexual populations, however, since a population which attains species-status may evolve specialisations for particular conditions, without interbreeding with other populations subjected to selection to suit different conditions. Speciation is, therefore, essential to the progress of rapid evolution in which valuable gene-combinations are not swamped by interbreeding with others.

Allopatric speciation

Two populations or species are *sympatric* if their areas of distribution overlap, and *allopatric* if they do not. Speciation is marked by *reproductive isolation*, or failure to interbreed, between the two newly-formed species. Whether or not such reproductive isolation can develop between sympatric populations is a matter of controversy. Some theorists argue that allopatry, or spatial isolation, is essential at some stage of the speciation process or else cross-breeding between the incipient species would swamp out any differences produced by disruptive selection pressures.

Previously sympatric populations of European animals are thought to have been separated into allopatric western and eastern populations by the southern expansion of ice during the last Ice Age. The separated populations would have been subjected to different selection pressures in the different environments. When the ice retreated and the separated populations met again, they had diverged into distinct species which neither interbred nor competed with one another. Thus the newts *Triturus cristatus*, from the west, and *T. marmoratus*, from the east, probably had a common ancestor before the Ice Age. Their distribution now overlaps in France. Such allopatric speciation must have many parallels in situations where populations of plants or animals become separated by geographical barriers, if only of a minor kind.

One interesting result of allopatric evolution is the occurrence of *ring-species*. A well-known instance is that of the herring-gull (*Larus argentatus*) and the lesser black-backed gull (*L. fuscus*). In western Europe, these behave as true species and do not interbreed, but races of *L. argentatus* in North America and Siberia connect it with west Siberian races of *L. fuscus*. Individuals from adjacent populations in the continuous series can interbreed; the series would be considered one species except that the overlapping 'ends' of the ring are reproductively isolated. The overlap probably occurred relatively recently, by which time the hindrance to gene-flow caused by the great distance involved had allowed the extremes of the population to diverge genetically. Had the divergence not been sufficient to cause reproductive isolation by that time, the extremes would have interbred and we would refer to only one species of gull.

The origin of reproductive isolation

Darwin considered that reproductive isolation arose merely as a by-product of evolutionary divergence. Alfred Russel Wallace, however, believed that the hybrids between two diverging populations would be less well-adapted than either parent and so be eliminated by natural selection. There seems to be some truth in both of these proposals.

The mechanisms which produce isolation are extremely varied. Plant populations in the process of speciating may flower at different times or develop incompatibility responses to each other's pollen. Animals possess subtle behavioural responses which may prevent inter-population pairings. When herring-gull and lesser black-backed gull eggs are exchanged, the young hatch and grow up behaving like their foster-parents. They mate with their adopted flock and produce fertile offspring. The usual reproductive isolation between the two sorts of gull must be entirely behavioural. In such behavioural reproductive isolation, it is often the female's choice of mate which provides the isolating factor. Female frogs and toads, for instance, will mate only with those males which produce an appropriate mating call. Small differences in the call-frequency, pulse rate, duration or rate of repetition may cause the female to reject the male's advances. This simple mechanism may reproductively isolate two diverging populations. Analysis by oscilloscope (an instrument which can analyse sound to give a visual display of its characteristics) of the sound of two species of the frog genus *Hyla* shows that the difference between the mating calls is pronounced only in regions where the two species are sympatric. In regions where the two species are normally allopatric, however, their calls become very similar: geographical separation makes the reproductive isolation unnecessary. Females could distinguish those calls whose oscillograms looked different but not those which looked similar (Figure 48).

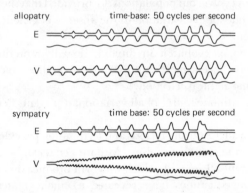

Figure 48   Oscilloscope-tracings of the mating-calls of *Hyla ewingi* (E) and *H. verreauxi* (V) from areas of allopatry and sympatry
Notice how similar are the traces from allopatric populations (which can never hear one another's calls) and how unlike are those from the sympatric populations (where confusion could occur if the calls were similar)

Such pre-mating reproductive isolation mechanisms are less wasteful of energy and gametes than mechanisms such as death or infertility of hybrid offspring and so will be promoted by natural selection.

Sympatric evolution

There is accumulating evidence that distinct races may occasionally arise sympatrically, that is, without any past or present geographical separation. One abrupt mechanism of sympatric speciation is that of polyploidy, discussed on page 61. Polyploidy apart, it has yet to be proved that sympatrically evolved races can evolve into true species.

Sympatric evolution of races requires a population to be divided by large contradictory selective forces. Good examples come from plants growing on mine dumps in soil contaminated with toxic heavy metals. The grass *Agrostis tenuis* growing on a Welsh lead-mine possesses an inherited tolerance to lead, whilst plants on an adjacent pasture, continuous with the mine, do not have this character. Powerful selective pressures against the non-lead-tolerant plants on contaminated soil maintain a boundary between the two sub-populations despite free pollen and seed dispersal between the two. The reasons for the disadvantage of the lead-tolerant plants on uncontaminated pasture are less clear.

SUMMARY OF THE THEORY OF EVOLUTION BY NATURAL SELECTION

Individuals do not evolve, but populations do, provided that they possess *heritable* variation. Mere phenotypic variation in the absence of genetic variety will not produce evolution.

The mechanism of evolution in diploid, sexually-reproducing organisms (those which have been most studied) requires three factors: mutation, recombination and natural selection.

Mutation is the ultimate source of all variation: it presents new possibilities to the organism.

Recombination of alleles during meiosis and fertilisation provides the immediate source of variation in sexual organisms. Mutant genes can be brought into the company of different alleles with which they will produce different phenotypic effects. Recent investigations have revealed a much greater proportion of heterozygous loci in many organisms than was suspected, all of them conferring genetic variety which may be recombined in various ways.

Natural selection 'chooses' the best-adapted organisms (that is, those with the most favourable collection of phenotypic characters) to survive and breed more successfully than their less well-endowed competitors. Natural selection has the paradoxical ability to provide phenotypic stability or diversity depending on the circumstances under which it acts. There is modern experimental evidence to show that the powers of natural selection in the wild are much greater than was once supposed.

Arguments against the theory

Opponents of the theory of evolution by natural selection (though not of the *fact* of evolution) make a number of criticisms:

1 Mutations are usually harmful, so it is impossible to conceive that random mutations could accumulate to the benefit of a species.

2 Whilst complex organs, such as the eye, function effectively, it is difficult to imagine how a half-evolved eye would be useful. (The theory of natural selection demands that each 'stage' in the evolution of an organ must be beneficial in itself, without the need to propose a directive force which 'works towards' a distant end.)

3 The theory does not account for the evolution of the higher taxonomic categories (genus, family, order, class, phylum).

These criticisms can be met as follows:

1 Genes do not act in isolation but as part of a gene-complex which buffers the effect of a particular gene. Natural selection acts on the gene-complex in such a way as to alter the phenotypic expression of a mutant gene and render its effects harmless. A laboratory example of this phenomenon involved the use of *Drosophila* carrying a mutation which reduced or eliminated the eye. Homozygous eyeless stocks of flies were allowed to inbreed for many generations with no artificial selection applied. Eventually, most of the descendants' eyes were normal. In each generation, selection had favoured those flies whose 'eyeless' genes were in the most favourable gene-complex until the harmful phenotypic effects were suppressed. These normal-looking flies (carrying two genes for eyelessness) were then crossed with wild-type flies to give heterozygotes with normal phenotypes. Selfing the heterozygotes gave the usual proportions of 1 dominant homozygote : 2 heterozygotes : 1 recessive homozygote in the F_2. This time, the homozygous recessives showed the original symptoms of eyelessness. The recombination of genes during outcrossing had stripped the 'eyeless' genes of their 'modifier' genes which reduced eyelessness. These modifiers had accumulated in the gene-complex by selection.

Since mutation is a rare event, organisms are heterozygous for a mutant gene when it first appears. Harmful effects can be suppressed if they are *made* recessive, and natural selection therefore favours those organisms in which the mutant gene's phenotypic effects are minimal, until total recessiveness is achieved.

All organisms carry a reservoir of potentially harmful mutant genes whose expression is limited or suppressed. The majority of these genes will have minor effects, since those with profoundly harmful attributes are likely to kill the organism in which they arise before natural selection has the opportunity to modify their effects.

Polygenic inheritance is particularly suited to subtle evolutionary change. Since a particular character is under the control of many genes, none of which has a very great effect, one mutation will not cause much phenotypic change. More likely, that character will merely appear a shade more extreme so that·a gradual shift occurs as mutations accumulate. Such a process is much less likely to cause dramatic, harmful effects than the mutation of a 'major', non-polygenic gene.

Suppressed mutant genes which are kept 'in reserve' may have effects which are beneficial under changed environmental circumstances. The genes conferring lead-tolerance to *Agrostis tenuis* appeared to 'served no useful purpose' until the grass encountered lead in its surroundings.

Mutation is a much more subtly-controlled event than was once supposed and is capable of supplying necessary variation, despite some initially harmful effects. As Sir Julian Huxley put it: 'The offer made by mutation to a species is not necessarily final. It may be a preliminary proposal, subject to negotiation.'

2 Complex organs develop their attributes gradually. A simple light-dark detector performs less well than a vertebrate compound eye, yet its possessor may

survive better than a competitor without it. Improvements *can* be added gradually.

The ability of certain fish to administer large electric shocks which stun their prey was once considered mysterious, since a mild shock does not have this effect. The mechanism is too complex to be the result of a single mutation and mild-shock intermediates were apparently useless. It was difficult to see how powerful shocks evolved without imagining a force which drove the evolution in a pre-destined manner. Lissmann solved the problem by demonstrating that mild shocks *do* have a function – but a different one: that of creating a 'radar-system' for detecting living prey. Presumably, progressively more efficient radar systems evolved to the point where the shock was sufficient to stun small prey. Undoubtedly, the same kind of reasoning could be applied to other mysteries if we were sufficiently knowledgeable about the intermediates.

3   The evolution of the higher taxonomic categories appears to require a different process from that of speciation. The discontinuities between the groups is wide; there are no organisms half-way between ferns and conifers or between butterflies and beetles.

The problem lessens when we realise that existing plants and animals are a mere fraction of those types which have existed. The extinction of a group's relatives leaves it distinct and becoming more distinctive as it diversifies and specialises. Fossils fill some gaps, although even these represent a biased sample of past life. Most organisms were not suitable for fossilisation and the earliest ones have been destroyed by crushing and heat in the rocks. If every organism that has lived could be resurrected, the collection would be much more of a continuum than a collection of all living species. Life has evolved over the last several million years. In such a time-span, events become inevitable which would be highly improbable in the short-term. (See also chapter 11.)

QUESTIONS

1   When DDT was first introduced as an insecticide, it successfully killed houseflies when used in low doses. Now, much greater concentrations are required and many fly populations are almost totally resistant to DDT. Explain this in terms of the theory of evolution by natural selection.

2   Explain what is meant by (a) allopatric and (b) sympatric evolution. Argue either for or against the proposal that two animal species may evolve sympatrically from one ancestor species.

3   Explain why diseases, such as cancer, which predominantly affect the elderly cannot be eliminated by natural selection.

4   Mimicry is the resemblance of one species to another to an extent which confuses potential predators. In Batesian mimicry, a *model* species, which is distasteful in some way to its predators, is resembled by inoffensive mimics of one or more different species. The mimics are avoided by predators which avoid the model. Explain how the mimicry may have come about. What balance must be maintained if the mimicry is to remain effective?

5   The banded snail (*Cepaea nemoralis*) is found with different colours and dark brown bands in different environments. The colour may vary from yellow (which appears as green when the snail is alive) to brown and pink. Over 200 separate investigations of the pattern of colour and banding related to the environment gave the following results:

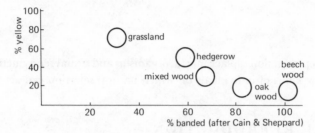

Discuss the factors which must balance heterozygosity in animal populations. how you might plan an investigation into your hypothesis.

6   In order to investigate the process of natural selection in a controlled environment, a plastic box was set up with jars of food media and a muslin-covered ventilation lid. Into the box were placed specimens of virgin female *Drosophila* of each of the following types:

wild type (red eye, long wing)
white eye, long wing
red eye, vestigial wing

Equal numbers of males of these three types were introduced at the same time and the whole environment was kept at an optimum temperature, at which the life cycle of the flies takes about three weeks.

Every three weeks, the populations in the boxes were etherised, counted, and then returned. The percentage composition of the population was recorded, this involving several thousand flies by the end of the experiment. Results were as follows:

	1st count (%)	2nd count (%)	3rd count (%)
Wild type	80.0	92.0	97.00
White eye, long wing	6.0	1.0	0.5
Red eye, vestigial wing	10.0	7.0	3.5
White eye, vestigial wing	4.0	0	0

Explain what this experiment shows about selection pressure against mutants and suggest possible biological disadvantages of the mutants used. Would the mutant genes soon be lost from the population?

If we assume that the white-eyed flies are blind, what sort of results might have been achieved if the experiment had been carried out in the dark? Discuss the factors which must balance heterozygosity in animal populations.

8 Reproduction

All forms of reproduction replace mortal organisms and sexual reproduction also generates new kinds of individuals on which natural selection can act.

ASEXUAL REPRODUCTION

Few organisms rely exclusively on asexual reproduction but many, particularly bacteria, protozoa and plants, reproduce asexually as well as sexually. Asexual reproduction depends on mitotic nuclear division.

Single-celled organisms such as bacteria, protozoa and one-celled algae simply divide by *binary fission* to produce two identical daughter cells. This process can proceed very rapidly. Certain single- and multicellular organisms convert cells into *spores*, each of which contains a set of genetic material and is adapted for dispersal. Many spores, such as those of fungi, are extremely durable and can be carried in air currents for long distances without losing the ability to germinate and grow once favourable conditions are encountered. Multicellular organisms, such as the pond *Hydra*, may grow offspring as outgrowths of their bodies in a process called *budding*. Some plants fragment in a reproductively comparable process: multicellular pieces may break off from a parent alga, for example, and start new plants.

The most highly-differentiated plants have a variety of specialised methods of *vegetative reproduction*. Modified stems may form above the ground (in which case they are called *runners* or *stolons*), as in the strawberry and bramble, or below it (when they are called *suckers*), as in mint or bindweed. In both cases, a young plantlet forms at the end of the stem and, once it has grown its own roots and leaves, may be severed from its parent to grow independently. Horizontal underground stems which store food and propagate the plant are called *rhizomes*. Some are fleshy, like those of the iris, while others (e.g. couchgrass) are wiry. Whole, thickened underground *stem-tubers* store food in plants such as potatoes and the tubers also serve to propagate the plant. Fleshy storage-roots behave in a similar way in lesser celandines, for example, in which small *root-tubers* readily break off and grow. *Bulbs* are underground buds which multiply vegetatively in plants such as lilies and onions. Some plants (e.g. *Begonia*) reproduce vegetatively with the aid of specialised leaves which grow plantlets spontaneously or when injured. In all these cases, a relatively large food-reserve from the parent plant is passed to the offspring, which therefore has a good chance of survival. Such methods of vegetative reproduction are suited to the colonisation of 'difficult' habitats (such as shifting sand) where the establishment of seedlings would be impossible.

Parthenogenesis is the development of a new organism from an unfertilised egg. In plants there is a variety of ways in which eggs may develop into embryos without fertilisation (although sometimes the plant must be pollinated merely to stimulate the embryo's growth). Dandelions produce their prolific seed asexually in this way. Certain animals, notably the small freshwater rotifers and aphids (greenfly), can reproduce parthenogenetically under favourable conditions. Aphids normally reproduce sexually at the end of the summer, so their's is a *facultative*, rather than an *obligate*, asexuality. In bees, parthenogenetic offspring are routinely produced by the queen; they are all male. Fertilised eggs give rise to females.

SEXUAL REPRODUCTION

In the life of a sexually-reproducing organism some cells are diploid and others haploid. Fusion of haploid gametes gives rise to the diploid phase (or *diplophase*) and meiosis of diploid cells produces the haploid phase (*haplophase*). Diplophase and haplophase alternate in the *life-cycle* of the organism.

Some unicellular organisms spend most of their life-cycle in the haplophase condition, the zygote being the only diploid cell, while others are always in diplophase with only the gametes being haploid. Multicellular animals are diploid, except for the gametes, and most species have two separate sexes. Multicellular plants have evolved a wide variety of life-cycles, some of which are highly complex. Most vascular plants produce male and female gametes in one body: they are *hermaphrodite*. Life-cycles in which both haplophase and diplophase have a multicellular body are said to have *alternation of generations*.

Alternation of generations

The sea-lettuce, *Ulva*, a marine alga, exhibits a simple alternation of generations. The thallus ('leaf') occurs in a diplophase and a haplophase, indistinguishable to the naked eye. The diploid thallus is known as the *sporophyte* because it produces cells which undergo meiosis to form haploid *spores*. Each spore swims with the aid of threads, called *flagella*, and eventually settles and multiplies to produce a haploid thallus. This thallus may be of either of two genetically-determined *mating-types*; it is called the *gametophyte* because it produces haploid gametes by mitosis. Two gametes of different mating-types may unite to form a diploid zygote which proliferates mitotically to give rise to the next sporophyte generation (Figure 49). The two generations are equally prominent in the life-cycle.

Mosses and liverworts, land plants without vascular tissue, have a life-cycle in which the haploid gametophyte predominates. Motile sperm swim through a film of water and fertilise eggs; each zygote so formed is the first cell of the sporophyte generation. The sporophyte remains attached to the gametophyte

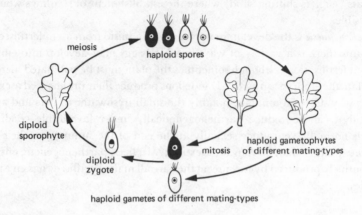

Figure 49   The life-cycle of *Ulva*, showing alternation of generations

and nutritionally dependent on it A sporogonium forms on a stalk of sporophyte tissue and, within it, haploid spores form following meiosis. These mature, are shed and germinate to give rise to the next gametophyte generation.

In the vascular land plants, the *ferns, gymnosperms* (mostly cone-bearers or *conifers*) and *flowering plants*, it is the diploid sporophyte which predominates. Ferns have small, free-living gametophytes but the gametophytes of gymnosperms and flowering-plants are produced within the sporophyte body. These latter two groups are *seed-plants*, the embryo of the new sporophyte generation being formed within a *seed* developed in the female gametophyte. (Since this gametophyte is borne within the parent sporophyte, superficially it appears that one diploid sporophyte directly gives rise to the next generation sporophyte without the intervention of a haploid gametophyte generation.) In seed-plants, the male gametophyte is conveyed to the female gametophyte as a *pollen-grain* which is impervious to water and so can travel through air without drying up. The evolution of the pollen-grain was an important step in the colonisation of dry land habitats by vascular plants.

REPRODUCTION IN FLOWERING PLANTS

In ferns, the leaves which bear spores (called *sporophylls*) are indistinguishable from ordinary vegetative leaves except for the presence of spore-bearing structures (sporangia) on their backs. Conifers have their sporophylls compactly grouped into cones which do not resemble vegetative leaves. Flowering-plants possess much-modified sporophylls in the form of *stamens* and *carpels*.

A stamen contains micro-sporangia, or *pollen-sacs*, which mature and rupture to release micro-spores, or pollen-grains. Within each grain, a minute male gametophyte develops.

A carpel is a hollow megasporophyll containing mega-spores, or *ovules*. Inside each ovule, a much-reduced female gametophyte or *embryo-sac* develops. Each

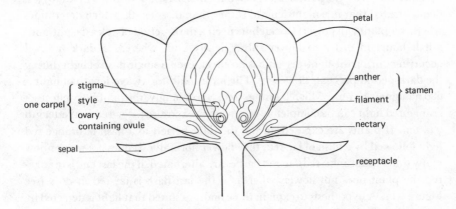

Figure 50   Half of a buttercup flower

group of stamens and carpels is surrounded by other modified leaves called *sepals* and *petals*; the whole structure is a *flower*.

Flower structure and function

The range of types of flower-structure is immense, although most hermaphrodite plants contain the same functional parts (Figure 50).

The axis of the flower is a short section of stem called the *receptacle*. From the receptacle extends the female part, the *gynaecium* which consists of one or more *carpels*. Each carpel swells into an *ovary* containing *ovules* (eggs) at the base, and tapers into a *style* whose tip is called the *stigma*. The stigma is the receptive surface for the pollen. Surrounding the gynaecium are the *stamens*, each consisting of a stalk or *filament* supporting a pollen-containing *anther*. The stamens together form the male part of the flower, the *androecium*. In flowers, these reproductive parts are surrounded by a ring of petals, often brightly-coloured, comprising the *corolla*, and a set of green *sepals*, collectively called a *calyx*. Flowers have evolved into a wide variety of forms. The arrangement in plants such as dandelion and cow-parsley; where many flower-heads are grouped together is called an *inflorescence*.

Control of flowering

A problem which has long intrigued botanists is: 'What induces a plant to switch from purely vegetative growth to flowering?' A plant must be mature to flower, but even then it does so only under favourable conditions of temperature and daylength or *photoperiod*. Some plants, such as gladiolus and clover, flower only when the days are long in summer, while others, like chrysanthemum and tobacco, are short-day plants. For generations, horticulturalists have artificially altered the daylength of glasshouse crops to produce flowers out-of-season.

Plants without leaves do not flower, even if the daylength is ideal, implying that flowering depends on some substance transported from the leaves.

Experiments on the American cocklebur (*Xanthium*), performed in the 1930s, demonstrated that it is night-length, rather than day-length, which determines whether a plant will flower. Cockleburs need a dark-period of at least eight-and-a-half hours in order to flower subsequently and, whereas a dark interval occurring during the light-period has no effect, even a single flash of light during the dark-period prevents flowering. The most inhibitory wave-length of light is 660 nanometres (nm), which is orange-red. Surprisingly, the inhibitory effects of orange-red light can be completely removed by exposure to far-red, wave-length 735 nm. If plants are exposed during the dark-period to a flash of orange-red light followed by a flash of far-red, they flower normally. It does not matter how many times the orange-red/far-red flashes are alternated, if the last one is orange-red the plant does not flower, whereas if the last flash is far-red, it does (see Figure 51). A hypothesis to explain these findings stated that light is detected by a pigment which exists in two forms, one absorbing mostly orange-red light and the other mostly far-red. Support for this hypothesis came later, with the isolation of the pigment *phytochrome* which has just these properties.

It seems that sunlight (containing orange-red light but little far-red) acts on the phytochrome and converts it all into the form sensitive to far-red light (called P_{735}). During the night, the P_{735} is converted slowly by enzymes into the orange-red sensitive form, P_{660}. The P_{735} form, produced by exposure to orange-red

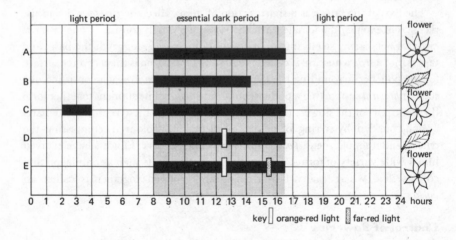

Figure 51   Control of flowering in the cocklebur
The shaded block represents the period of eight-and-a-half hours' darkness, normally essential for cockleburs to flower. Black bars show how much time the plant actually spent in darkness during each trial. Flowering occurred in trials A ($8\frac{1}{2}$ hours' darkness) and C ($8\frac{1}{2}$ hours' darkness plus an extra 2 hours) but not B (less than $8\frac{1}{2}$ hours' darkness). If an $8\frac{1}{2}$-hour dark period was interrupted by a flash of orange-red light, no flowering occurred (D), but a flash of orange-red followed by far-red light allowed the plant to flower normally (E)

light, is the one which inhibits flowering, and even a flash of orange-red is sufficient to convert the non-inhibitory P_{660} back to P_{735}. The dark-period necessary to flowering was once thought to be measured by the plant as the time taken to convert phytochrome from the inhibitory to the non-inhibitory form. Now we know that the conversion takes only a fraction of the required dark-period; the remaining time must be used for other preparations, probably the synthesis of the flowering-hormone in the presence of the non-inhibitory pigment. This flowering-hormone is then transported to the shoot-apex from the leaves, and induces the development of the floral parts.

The anther and pollen-grain production

The young anther develops four lobes, each containing a vertical row of enlarged cells. Each large cell divides to form an inner *sporogenous cell* and an outer *parietal cell*. The parietal cell divides rapidly, producing a layer, several cells thick, around the sporogenous cell. The innermost row of cells develops into the *tapetum* or nutritive layer. Meanwhile, the sporogenous cell multiplies repeatedly to produce a mass of *pollen mother-cells* contained within a space called the sporangium or pollen-sac (Figure 52 (a)). Up to this point, cell-divisions have followed mitotic nuclear-divisions, but now each mature pollen mother-cell undergoes meiosis and divides into a tetrad of four haploid cells. These

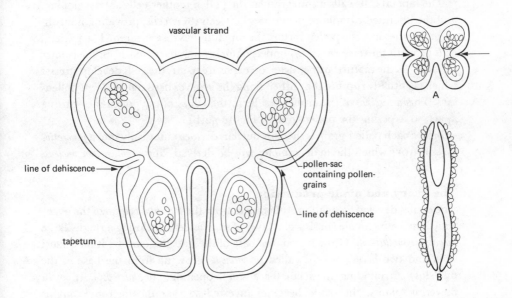

Figure 52(a) (left)   Transverse section (T.S.) through an almost-ripe lily anther to show the pollen sacs (× 20)
(b) (right)   T.S. through tulip anther before (A) and after (B) dehiscence (× 4). The anther ruptures along its length at the grooves marked by arrows, so that the two halves spring apart, exposing the pollen

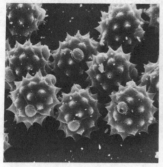
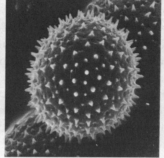
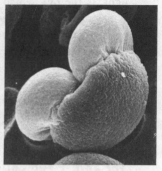

Chrysanthemum myconis   Malva sylvestris   Pinus sylvestris

Figure 53   Grains of pollen from insect- and wind-pollinated flowers

represent the spores of the plant's sporophyte generation. Each matures into a *pollen-grain*, surrounded by a thin membrane called the *intine* and a thick *extine*. which is variously sculptured according to species (Figure 53).

The tapetal cell-walls disintegrate as the pollen mother-cells are developing, and the liberated cytoplasm penetrates between these cells, providing nourishment. By the time the pollen is ripe, the tapetum has disappeared. The parietal cells beneath the *epidermis* (skin) undergo thickening, particularly on their inner surfaces. As the mature anther dries, on exposure to air, the cells develop stresses which eventually rupture the anther along the grooves between adjacent pollen-sacs. This *dehiscence* of the anther results in the edges of the pollen-sacs curling apart, so exposing the pollen-grains (Figure 52(b)).

Before each pollen-grain is shed, its nucleus divides mitotically into a *generative nucleus* (from which the male gametes will be derived) and a *pollen-tube nucleus*.

The ovary and ovule-production

The ovules develop as small protuberances from the *placental* region of the ovary wall. The young ovule consists of a layer of *nucellus* cells enclosing a single, large *megaspore mother-cell*. Growth from the base of the nucellus elevates it on a short stalk and two layers of cells called *integuments* grow up from the base of the nucellus, surrounding it except for a small opening, the *micropyle*. In most flowering plants, the ovule becomes inverted so that the micropyle points towards the placenta and the integuments and stalk fuse. Following two meiotic nuclear divisions, the megaspore mother-cell divides twice to form four megaspores. Three megaspores disintegrate and the fourth gives rise to the female gametophyte, known as the *embryo-sac* (Figure 54).

Embryo-sac formation

The surviving megaspore nucleus undergoes three mitotic divisions to give eight

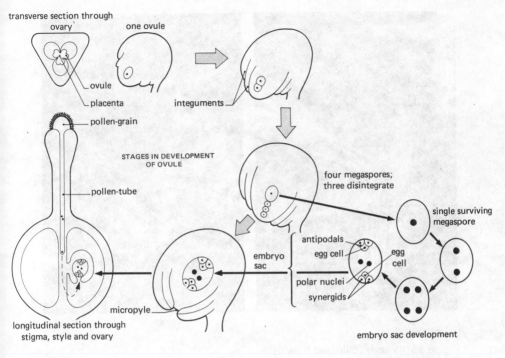

transverse section through ovary

one ovule

ovule

placenta

integuments

pollen-grain

STAGES IN DEVELOPMENT OF OVULE

four megaspores; three disintegrate

pollen-tube

single surviving megaspore

antipodals

embryo sac

egg cell

egg cell

polar nuclei

synergids

micropyle

longitudinal section through stigma, style and ovary

embryo sac development

Figure 54   Highly diagrammatic representation of the development of an ovule until the time of fertilisation

genetically-identical nuclei. As the ovule absorbs food and water through its stalk, the embryo-sac, nucellus and integuments enlarge. Four nuclei cluster at the micropyle end of the sac and four at the opposite end. A single *polar* nucleus then moves from each cluster to the centre and the remaining nuclei become invested in membranes. At the micropyle end, the three cells are called the *egg-cell* and two *synergids*, while the three cells at the opposite end are the *antipodals*. After the formation of these cells the ovule is mature (Figure 54).

Pollination mechanisms

Most flowers are organised in such a way that transfer of pollen from one flower to another (as well as merely from anther to stigma) is facilitated. Many plants possess mechanisms which ensure that cross-pollination occurs (that is, that the flowers involved are on separate plants). The usual agents for conveying pollen from one plant to another are wind and insects, although water and some birds and bats also have rôles as pollinators.

Wind-pollination

The advantage to the plant of *wind-pollination* is that it is independent of the occurrence of insects. Huge quantities of pollen must be disseminated to make pollination reasonably certain – about a million grains to the square metre of

Figure 55   Wind-pollinated flowers
(left) Spikelet of false oat, *Arrhenatherum elatius*
(right) Inflorescence of ribwort plantain, *Plantago lanceolata*

plant habitat are necessary if there is to be a reasonable certainty of one falling on a 1 mm² stigma. To meet this need, one birch catkin, for example, releases about $5\frac{1}{2}$ million pollen-grains.

Wind-pollinated plants, which include most British trees and virtually all grasses, produce smooth, dry, light pollen-grains, sometimes with wings or air-bladders. The anthers are large and exposed, often dangling from long filaments or in catkins, while the stigmas are projecting and feathery. These adaptations facilitate dissemination and capture of pollen respectively. It is usual for wind-pollinated plants to have separate-sex flowers (or else for the stamen and stigma to ripen at different times) or else the stigma would become totally clogged with its own flower's prolific pollen. The petals and sepals are reduced and inconspicuous and do not interfere with the free-flow of pollen (Figure 55).

Insect-pollination
Insects are attracted to insect-pollinated flowers by scent, petal-colour and reflections of ultra-violet light. Once the insect arrives at the flower, it can obtain pollen and *nectar* (a sugary liquid produced in pouches called *nectaries*).

The pollen of insect-pollinated flowers is heavier than wind-pollen, with a sticky or spiky surface which clings well to insect's bodies. It is produced less abundantly than wind-pollen as it is more likely to reach a stigma.

Insect-pollinated plants and their pollinating insects exert powerful selective

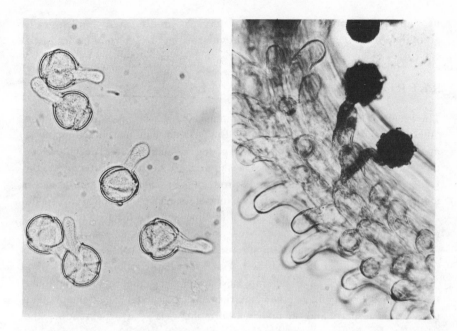

Figure 56   (left) Pollen grains from stinking hellebore (*Helleborus foetidus*)
germinating in a 10 % sugar solution, showing pollen tubes
(right) Germinating pollen grains on the stigma of chickweed, *Stellaria media*

pressures on one another. Such a close relationship results in a mutually
dependent *co-evolution*.

A variety of examples of pollination mechanisms is shown in chapter 17 of *The
diversity of life*.

Germination of pollen

When a pollen-grain lands on the ripe stigma of a plant of the same species, it will
germinate, provided that there is no incompatibility reaction between the grain
and the stigma (see p. 85). Germination entails the rupturing of the grain and the
growing of the intine through a pit in the extine to form a *pollen-tube*. The tube
penetrates the stigma, nourished by sugary fluid secreted there, and grows down
the style (Figure 56). If the generative cell has not already divided, it does so in
the pollen-tube to form two male gametes which move down the tube. The end of
the tube penetrates the embryo-sac (usually via the micropyle) and bursts,
releasing the gametes.

Fertilisation

One of the two gametes fuses with the egg-nucleus to form the zygote from
which the embryo plant will develop. The other gamete joins with the fused
polar nuclei (i.e. the *primary endosperm nucleus*) to form the *secondary endosperm
nucleus* which is triploid. This *double-fertilisation* is unique to flowering-plants;

its evolutionary origins remain a mystery. Normally, many pollen-grains germinate on a stigma so that many ovules in one ovary may be fertilised.

REPRODUCTION IN ANIMALS

The great majority of multicellular animals reproduce only sexually. Unless otherwise mentioned, descriptions in this section will be limited to those mammals which are *placental*, that is, the young are nourished and develop inside the mother's body, attached to a *placenta*, and are born live. After birth, the young feed from the mother's milk-producing or *mammary* glands. Such a method of reproduction gives a good chance of survival to each offspring, so few offspring are produced in the parent's lifetime and the litter-size at birth is relatively small.

In this chapter, the mechanism of reproduction will be considered, while the development of the embryo will be reserved for chapter 9.

Mammals always have separate sexes possessing different reproductive systems. The sex-organs, or *gonads*, are the *testis* (male) and *ovary* (female). The testes produce sperm and the male sex-hormones (*androgens*) which induce and maintain the characteristics of maleness. The ovaries produce ova (eggs) and female sex-hormones (*oestrogens*) which cause the development of the female characters. Accessory organs are associated with the gonads; in the male these conduct the sperm to the outside, and in the female they receive the sperm and maintain the developing embryo if conception occurs.

Gametogenesis

Gametogenesis (gamete-production) is called *oogenesis* (egg-production) in females and *spermatogenesis* (sperm-production) in males. It is fundamental to sexual reproduction as it includes the essential meiotic divisions which halve the chromosome numbers of the prospective gametes. In both male and female animals, the *germinal cells*, those destined to become gametes, go through three stages which take place in the gonads:

1 Increase in numbers of cells, involving mitoses.

2 Growth. At the end of this stage, each cell is called a *primary oocyte* (female) or *primary spermatocyte* (male).

3 Maturation, including meiosis.

The cytoplasm of the male cells divides equally following the meiotic chromosomal divisions to give four sperm. After meiosis in the female, however, the cytoplasm concentrates around one of the four nuclei only. After the first meiotic division, a *secondary oocyte* and a tiny *first polar body* are formed, with equal amounts of genetic material. The second meiotic division produces an ovum with almost all the original cytoplasm, and a genetically-equivalent *second polar body*. (The first polar body may also undergo the second meiotic division at the same time as the secondary oocyte.) The polar bodies perish, so that only one quarter of the available genetic material appears in mature ova. Probably, the value of

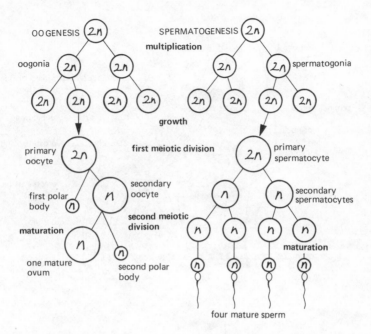

Figure 57 Gametogenesis
The stages in oogenesis and spermatogenesis are comparable, although oogenesis takes longer. Only one mature ovum derives from one primary oocyte, but a primary spermatocyte gives rise to four sperm

this unequal division of cytoplasm is that it gives the ovum a better chance of survival with its extra quantity of nutritive cytoplasm (Figure 57).

Gametogenesis takes longer in females than males. Sperm-production is continuous in mature males, but females are born with all the oocytes that they will shed during their lifetime already in their ovaries. In mammals, oocytes proceed to metaphase of the second meiotic division before ovulation and do not complete this division until after fertilisation.

The reproductive system of humans

The male
The two testes develop within the abdomen of the male embryo. Each becomes attached to a cord, the *gubernaculum*, which guides the testis into a pouch of skin, the *scrotum*, outside the abdomen usually before birth. The testis receives its blood supply via an artery and vein which run in the *spermatic cord*. Development of the sperm proceeds normally only at the temperature of the scrotum; in humans this is about 2°C cooler than that of the abdomen (Figure 58).

Testis structure and sperm-production
Each testis is composed of compartments containing convoluted *seminiferous tubules* where spermatogenesis proceeds. A cross-section through a seminiferous

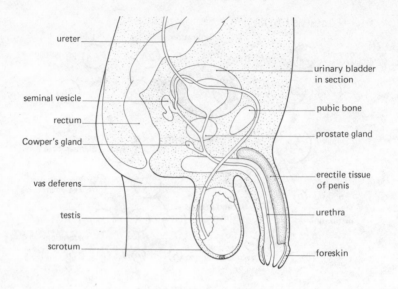

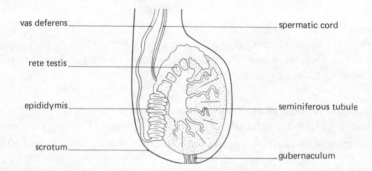

Figure 58(a)   L.S. and side view of the internal structure of the human male reproductive system
(b)   Human testis dissected to show details of sperm tubes

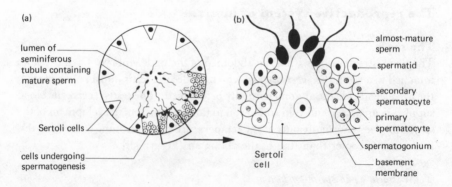

Figure 59   T.S. human seminiferous tubule
The section marked in (a) is shown enlarged in diagram (b)

tubule reveals large numbers of cells at various stages of multiplication and growth. The youngest cells, *spermatogonia*, are those proliferating from the epithelium of the tubule wall, while the more mature migrate into the lumen (cavity) of the tubule. When meiosis and cytoplasmic division are complete, the rounded haploid cells are called *spermatids*. These attach to pyramidal nutritive *Sertoli cells* where they mature into spermatozoa (sperm-cells). Sperm consists of little more than a DNA-packed head and a propulsive tail, and they are produced by the hundred million (Figure 59).

In each testis, mature sperm pass from the seminiferous tubules into the *rete testis*, a series of channels which lead to a single convoluted duct called the *epididymis*. Ciliated cells lining the epididymis tubules waft the sperm into the single duct which drains each testis, the *vas deferens* (Figure 58).

Release of sperm
Sperm-production continues whether or not the sperm are *ejaculated* (expelled to the outside of the body); if they are not, the vas deferens probably re-absorbs them. When ejaculation occurs, the sperm are conveyed to the *urethra* (a tube from the base of the bladder) by muscular contraction of the vas deferens. The sperm pass ducts leading from a *seminal vesicle* to each vas deferens. These glands secrete a thick fluid which dilutes and transports the sperm. Surrounding the widest part of the urethra is the *prostate gland*, and further along is a pair of *Cowper's glands*. These three glands contribute their secretions to the sperm. The mixture of sperm and fluids is called *semen*. The urethra conveys the semen out of the body through the *penis*. During sexual excitement, the many cavities in the penis become engorged with blood so that it enlarges and becomes rigidly erect; in this condition it can be inserted into the vagina of the female. Friction of the penis triggers a spurting of semen from the urethra by a series of muscular contractions, after which the penis returns to normal. The semen is deposited deep in the vagina and the sperm make their way by movements of their tails into the neck, or *cervix*, of the uterus (womb). Contractions of the uterus wall then waft some of the sperm into the Fallopian tubes where fertilisation may occur. In humans, about 550 million sperm are released during each ejaculation, although the vast proportion never come within reach of an ovum.

The female

Ovary structure and function
The female's two ovaries lie within the abdomen. The outer surface of each ovary is composed of a *germinal epithelium* whose cells proliferate to give rise to the germinal cells. In the embryo, clusters of cells migrate into the ovary tissue from the germinal epithelium; each cluster is called a *primary follicle*. At birth, the ovaries of a human girl contain about 70 000 such follicles, most of which eventually disintegrate and disappear, while a few mature and produce ripe ova.

Maturation of follicles
One of the primary follicle cells, the *oogonium*, larger than the rest, is destined to

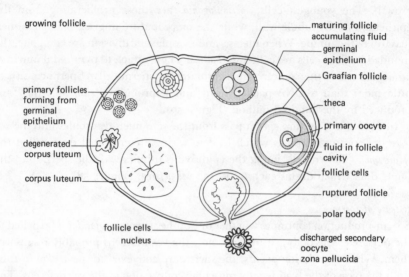

Figure 60   Diagrammatic representation of the stages in the development of an egg within the human ovary

become the ovum. At sexual maturity, the surrounding smaller *follicle cells* multiply to produce a layer several cells deep. A shiny layer called the *zona pellucida* appears around the developing egg. The follicle cells secrete a fluid which begins to accumulate between them. At this stage, the follicle is called a *Graafian follicle*.

The Graafian follicle continues to accumulate fluid and grows in size. The developing egg, now a primary oocyte, is pushed to one side where it attaches to the follicle wall. Connective tissue surrounds the growing follicle as a distinct double layer called the *theca*. The ripe Graafian follicle bulges from the surface of the ovary and eventually it ruptures, shedding the oocyte, surrounded by a halo of follicle cells; this is *ovulation*. By this stage, the oocyte chromosomes have undergone their first meiotic division and the cell is a secondary oocyte. The space left in the ruptured follicle eventually fills with large cells containing a yellow pigment which forms a structure called the *corpus luteum* (yellow body). The corpus luteum produces a hormone, *progesterone*, which prepares the uterus to receive a fertilised ovum (Figure 60).

The fate of the oocyte
The oocyte which breaks free from the surface of the ovary is collected by the funnel of the nearest *Fallopian tube* or oviduct whose fringed surface touches the ovary. The oocyte matures to become an ovum. Should intercourse have occurred, the sperm must reach the ovum in the Fallopian tube in order to fertilise it. Ciliated cells lining the Fallopian tube waft the fertilised ovum, dividing as it goes, towards the *uterus* (womb) (Figure 61). The ovum implants

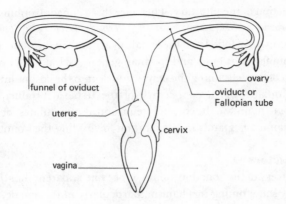

Figure 61 Front view of the internal structure of the human female reproductive system

into the prepared uterus-lining where it develops into an embryo. If the ovum is not fertilised, it passes out of the body via the vagina.

Hormonal control of reproduction

In addition to producing gametes, it has long been known that the gonads function as *endocrine organs*, that is, they produce *sex-hormones*, chemical messengers carried in blood. *Castration*, or the removal of the testes, of an immature animal prevents it from developing male *secondary sexual characteristics* (such as the deepening of the voice, the increase in strength and the development of face and body hair in young men) and ovary-removal has the equivalent effect on a female. Information on the chemical composition of sex-hormones is in chapter 3 of *The cell concept*.

In 1848, Berthold conducted a series of experiments on young cockerels. Castrated young cockerels failed to produce the wattle, comb or spurs characteristic of a mature animal when they grew up. If these castrated birds received a transplant of a testis from a mature animal, they did develop male sexual characters. The implanted testes developed blood connections with the cockerel's body, but no nervous connections. This demonstrated that the effect of the testis on the cockerel's development is a hormonal one.

Gonadotrophins

In the 1920s and 30s, investigations began to elucidate the biochemistry of the sex-hormones and to discover what dictated their production. Control was found to come from the *pituitary gland* and its removal produces in animals a regression of reprductive functions comparable with that produced by castration or ovary-removal. The pituitary is a structure about 1 cm across in humans, connected by a stalk to the base of the brain. This master-gland issues 'instructions' in the form of hormones that regulate the activity of all the body's other endocrine glands. Hormones which stimulate the gonads are called *gonadotrophins* and come from the anterior section of the pituitary. There are three gonadotrophins, the *follicle-stimulating hormone* (*FSH*), the *luteinising hormone* (*LH*) and *prolactin*. FSH promotes the growth of follicles in the female ovary and, in the male, it stimulates growth of seminiferous tubules and spermatogenesis. LH induces ovulation and the conversion of the cells lining the empty follicle into the corpus luteum. In the male, LH stimulates the *interstitial cells* of the testis to produce androgens. Prolactin helps to maintain the corpus luteum and also stimulates the secretion of milk in the mammary glands in preparation for feeding the young.

Releasing factors

After the discovery of the gonadotrophins, it became apparent that their activity is regulated by some timing mechanism, particularly in the female, where egg-production is cyclical. External events were thought to provide the cues for a biological 'clock' which triggers gonadotrophin production. The nervous system is apparently involved in such control, at least in mammals such as the ferret, since the day-length, perceptible only through the eyes and brain, affects reproductive hormones. The problem was to explain how the brain might influence the anterior pituitary when there were known to be no nervous connections between them. The answer was provided in 1947, when it was demonstrated that a blood-vessel connection links part of the brain (the *hypothalamus*) with the anterior pituitary. Nerve fibres from the hypothalamus terminate on the walls of blood-capillaries and secrete chemicals into them, by a process known as *neurosecretion*. The chemicals are called *releasing-factors*, a special kind of hormone with relatively simple structure. The capillaries drain into a vein which enters the anterior pituitary and breaks up into another capillary system (Figure 62). (Such a circuit, with capillaries at both ends, is called a *portal system*.) There is, therefore, a direct route for releasing-factors to be transported from the hypothalamus to the pituitary without being diluted in the general blood-circulation. The releasing-factors are produced in response to instructions

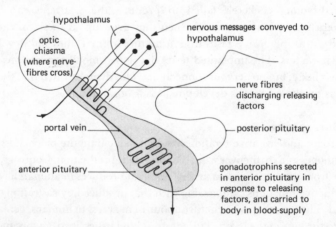

Figure 62   Diagram of the events in the anterior pituitary leading to the secretion of gonadotrophins

from the hypothalamus and they stimulate the pituitary to secrete gonado-trophins. Since the hypothalamus is directly connected to other brain-centres, there is an obvious mechanism by which the perception of environmental and psychological factors may affect gonadotrophin production.

Control of male reproduction

In male mammals, FSH is produced continuously so that sperm-production is uninterrupted. LH stimulates androgen-production in the testis until the androgens in the blood reach a certain level which inhibits the pituitary from producing more LH. The fall in LH levels causes the interstitial cells to stop producing androgens, so that the pituitary is eventually released from its inhibition and secretes LH again. Androgen-production is therefore controlled by a feedback mechanism. FSH and LH act *synergistically* i.e. the two hormones together have a greater effect than the sum of the contribution of each in isolation. LH has the dominant rôle once spermatogenesis is established.

Control of female reproduction

Most female mammals mate only during the breeding season when they experience periods of sexual activity called *oestrus* ('heat'). These periods alternate with inactivity (*anoestrus*). Some animals, including rodents and primates (the group to which monkeys, apes and man belong), experience oestrus cycles all the year round. Women do not show periodic oestrus; instead they undergo *menstrual cycles*, which are marked by the monthly shedding of the uterus lining with some blood, an event called *menstruation*.

The oestrus cycle

In response to the appropriate releasing-factors, the pituitary secretes FSH and small quantities of LH which activate the ovary. One (or more) Graafian follicle develops and, as it does so, it secretes oestrogen into the blood. The oestrogen stimulates the thickening of the uterus lining, increasing its blood-supply and prompts the development of ducts in the mammary glands. At certain oestrogen levels, the mating-behaviour characteristic of oestrus is induced. As the oestrogen-level increases, it inhibits the pituitary's FSH production and the oestrogen surge is followed by a rapid surge of LH and one of FSH. After peak levels of LH are reached, ovulation and the development of the corpus luteum occur (so cutting off oestrogen production). As it grows, the corpus luteum secretes increasing quantities of progesterone. Under progesterone's influence, the thickness and vascularity of the uterus continue to increase and the mammary glands develop tissue for the secretion of milk. Eventually, the corpus luteum gradually degenerates, lowering the progesterone levels, and levels of FSH and LH begin to rise to start a new cycle (Figure 63).

The stages in the oestrus cycle can be recognised by taking vaginal smears, since the vagina undergoes a characteristic sequence of changes. Observations are most easily made on small rodents such as mice, in which the peak of oestrus is characterised by the presence of many thin, flat cells. White blood-cells appear in smears taken at the end of oestrus.

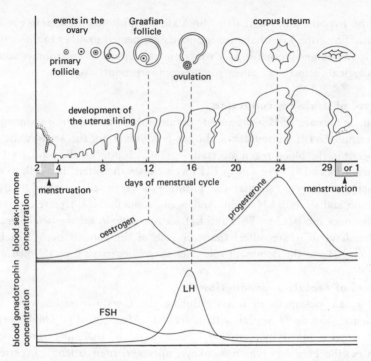

Figure 63   The relationship between the concentration of ovarian hormones and gonadotrophins, the events in the ovary and the development of the uterus lining in a woman

The cyclical rhythm is maintained by dynamic interrelationship between the sex-hormones and the gonadotrophins; the mechanisms of control are complex. The effect of this elaborate mechanism is to bring the uterus into a state of readiness for implantation of the embryo, should fertilisation occur. In this case, the presence of the embryo halts further oestrus cycles until after the birth.

Menstruation
In women, there is a total breakdown of the thickened *endometrium* (inner lining of the uterus) after the disintegration of the corpus luteum. The dead tissue is shed and discharged from the vagina, along with blood and the unfertilised ovum; this constitutes menstruation.

Pregnancy
Should fertilisation of the ovum occur, the corpus luteum continues to function. Following fertilisation in humans, therefore, menstruation (which would sweep away the fertilised ovum as effectively as it would an unfertilised one) does not follow ovulation.

The signal which causes the maintenance of the corpus luteum is a gonadotrophic hormone emitted by the fertilised ovum once embedded in the uterine wall. It is vital that the hormonal signal from this minute quantity of tissue is powerful enough to effect the corpus luteum or the pregnancy will fail from the outset. Normally this *chorionic* gonadotrophin can be detected in the blood of a pregnant woman a few days after implantation of the embryo. Its presence is the basis for most pregnancy tests.

As pregnancy proceeds, the developing placenta continues to produce gonadotrophin, as well as its own oestrogen and progesterone, which promotes further growth of the uterus, vagina and mammary glands. These hormones also inhibit pituitary production of FSH and LH and hence prevent follicle-growth and ovulation. In humans, but not in rodents, the corpus luteum eventually becomes redundant as a source of sex-hormones so that the ovaries may be removed without disturbing the pregnancy.

The contraceptive pill
Ovulation can be suppressed by 'fooling' the pituitary with large doses of oestrogen and/or progesterone taken by mouth. These hormones are constituents of the various sorts of contraceptive pill and they simulate the hormonal condition of pregnancy.

Length of reproductive life
In humans, *puberty* (the beginning of sexual maturity) is initiated by the gradual maturation of the gonads which start to produce the sex-hormones that induce development of the secondary sexual characters. Later, viable gametes are produced. In men, sperm-production usually continues from between the ages of thirteen to seventeen until old age. Ovulation begins in girls generally between twelve and fifteen years of age and comes to an erratic halt in middle-age. The complete cessation of ovulation is called *menopause*; it may take many months and is accompanied by hormonal disturbance.

Reproductive behaviour

The simplest method of bringing eggs and sperm together for fertilisation is to release them together into water. Since the gametes do not live long outside the body, it is essential that they are shed concurrently. Environmental factors, such as particular day-length, set off *spawning* (egg- and sperm-release) and frequently one individual then sets off another by secreting specific chemicals into the water. Gametes are liberated in vast quantities because the chances of fertilisation are low.

Internal fertilisation, or the direct transfer of sperm into the female reproductive tract, is essential to completely terrestrial animals, since gametes cannot stand exposure to air. In mammals, the penis is specialised as an organ for transferring sperm.

Courtship behaviour usually occurs between two animals before mating can take place. Sometimes this is elaborate, and it may involve the formation of a relationship (*pair-bonding*) which will last for some time – in the case of birds, long enough to construct a nest, lay eggs and rear the young.

(The remainder of this chapter refers to a variety of animal species.)

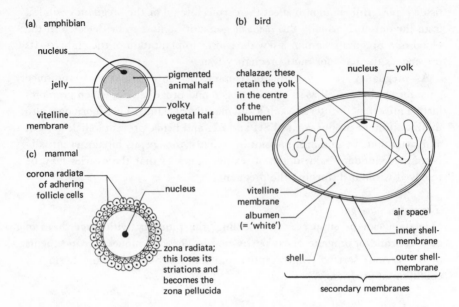

(a) amphibian

nucleus

jelly

vitelline membrane

pigmented animal half

yolky vegetal half

(b) bird

chalazae; these retain the yolk in the centre of the albumen

nucleus   yolk

vitelline membrane

albumen (= 'white')

shell

air space

inner shell-membrane

outer shell-membrane

secondary membranes

(c) mammal

corona radiata of adhering follicle cells

nucleus

zona radiata; this loses its striations and becomes the zona pellucida

Figure 64   Unfertilised eggs of amphibian, bird and mammal (Not drawn to the same scale)

Animal eggs

Mature ova, or eggs, can exceed a centimetre in diameter, while sperm are usually only a few micrometres long (excluding the tail). Eggs are non-motile, and their large size results from an accumulation of *yolk*, an energy-rich food. Yolk is a complex mixture of fats and proteins; in amphibians and birds, it is produced in the maternal liver and transported in the blood to the developing oocyte in the ovary. Different species vary greatly in the amount of yolk their eggs contain. The eggs of animals, such as fish, amphibia and birds, which must be nutritionally self-sufficient from the time of laying to the hatching of the young animal, contain a lot of yolk. Mammalian eggs, which are nourished throughout development via their placental connection with the mother, contain little yolk. Certain marine invertebrates' eggs (such as those of the sea-urchin) hatch rapidly into larvae which feed voraciously and these eggs, too, are laid with little yolk (Figure 64).

Yolk-distribution

The distribution of yolk within the egg-cell varies between species. Generally, one end of the egg is relatively yolk-free; this is called the *animal pole*. The other end, where yolk is concentrated, is the *vegetal pole*. Since yolk is denser than cytoplasm, aquatic eggs float vegetal pole downwards. Some eggs with relatively little yolk do not have a clear yolk gradient. The distribution of yolk profoundly affects the division of the egg after fertilisation.

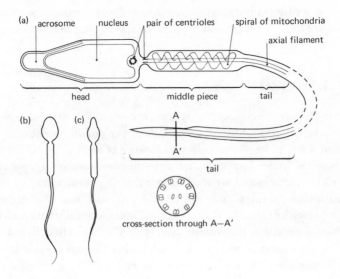

Figure 65   Diagrams of mammalian sperm
The head is viewed from the flattened side in (a) and (c) and from the broad side in (b). The majority of the tail is omitted in (a)

Egg-membranes

Eggs possess protective membranes of two different origins. *Primary membranes* are the *plasmalemma*, or cell-membrane, of the egg itself and the *vitelline membrane* which, in mammalian öocytes, arises as a result of interaction between the follicle cells and the öocyte, and becomes the zona pellucida. *Secondary membranes* are added to the eggs of some species as they pass down the oviduct; these include the 'white', the shell-membranes and the shell of birds' eggs, and the jelly of frogs' eggs.

Eggs of a variety of species can be induced to develop parthenogenetically into complete individuals. They must, therefore, contain all the information and materials necessary for development; they are the only cells in the bodies of most adult animals normally capable of such development.

Sperm

The sperm performs two functions: it activates the egg, initiating development, and it contributes a haploid set of chromosomes. Its life is short and its structure limited to the barest essentials, that of a self-propelled nucleus.

A mature sperm consists of three sections. The *head* is largely composed of a DNA-packed nucleus, topped by an *acrosome* which enables the sperm to penetrate the egg-membranes. The *middle-piece* contains mitochondria, the site of aerobic respiration; here, the energy necessary for sperm-movement is generated. A pair of centrioles nestle in a depression behind the nucleus and from one of them arises the axial filament of the *tail* or *flagellum*. This filament is composed of a ring of nine longitudinal double fibres surrounding two central fibres (Figure 65). The flagellum propels the sperm once it is liberated.

FERTILISATION

In animals, there is no evidence that sperm are chemically attracted to eggs. The vast numbers of sperm released (40–50 million per millilitre of human semen) ensure that a few come within collision-distance of an egg.

In 1919, F. R. Lillie observed that, if unfertilised sea-urchin eggs stand in sea-water for a few minutes and are then removed, the 'egg-water' has the ability to make sea-urchin sperm stick together. He proposed that the egg emits a substance (which he called *fertilisin*) that combines specifically with molecules of *antifertilisin* from sperm of the same species. Molecules of fertilisin, free in the egg-water, effectively fasten sperm together but, in their natural position in the egg-jelly, fertilisin molecules attach sperm to the egg.

The acrosome- and cortical-reactions

Once the sperm's acrosome contacts the egg-surface, the outer acrosome membrane breaks down. An enzyme within the acrosome is released and begins to digest the vitelline-membrane. Meanwhile, the inner acrosome-surface develops filaments which migrate inwards. When the acrosome-filaments touch the egg's plasmalemma, they set off a sequence of events called the *cortical-reaction* of the egg (the *cortex* is the egg's outer layer). *Cortical granules* beneath the plasmalemma rupture, beginning at the point of sperm-entry and spreading around the cortex. Substances liberated from the granules spread beneath the vitelline membrane, thicken it and lift it clear of the plasmalemma. The vitelline membrane is now called the *fertilisation membrane*. For many years, it was supposed that it was these changes which made the egg impenetrable by further sperm once the first one had made contact. (It is a fact that, despite the number of sperm surrounding an egg, *polyspermy*, or the entry of more than one sperm, is very rare except in eggs with a high yolk content.) The wave of change which lifts the fertilisation membrane takes several seconds to encompass the egg, however, which seems to be too slow to have the necessary effect. It seems likely that some substance is released rapidly, which prevents adherence of subsequent sperm, but the matter is not yet fully resolved.

Sperm-entry

In some species, the egg-cytoplasm surges upwards as a *fertilisation-cone* which engulfs the sperm and withdraws it into the egg. In mammals, the whole sperm enters, whereas, in sea-urchins, the tail remains outside. If the surface cytoplasm is pigmented, as in amphibians, the path of the sperm inwards is marked by a pigment-trail. If the egg is of a type which does not complete meiosis until after sperm-entry, the sperm remains quiescent until this has been accomplished. The sperm and egg nuclei then fuse togther. In eggs which are entered by more than one sperm, only one male nucleus fuses with the egg-nucleus and the other sperm disintegrate.

The entry of a sperm causes rearrangements of cytoplasmic substances which are particularly easy to observe in frog eggs. Unfertilised frog eggs are deeply pigmented in the hemisphere nearest to the animal pole. A few minutes after sperm-penetration, the cortical (surface) cytoplasm streams upwards, towards the animal pole, on the side that will develop into the embryo's dorsal surface. On the future ventral surface, the cortical cytoplasm moves downwards towards the vegetal pole. A lightly-pigmented *grey crescent* marks the future dorsal surface. Even before the zygote first divides, then, it acquires a symmetrical organisation. It seems that the orientation of the future embryo depends largely on the arrangement of cytoplasmic substances, particularly those in the cortex. In many eggs, if the cortex is disrupted, development cannot proceed normally.

Some eggs can be stimulated to begin development without fertilisation if they are treated with various chemicals or given temperature shocks. This *artificial parthenogenesis* apparently works only because it damages the cortical granules and, once they have liberated their contents, the train of events leading to development is set off.

QUESTIONS

1 Discuss the events which lead up to the shedding of an egg from the ovary of a woman and the maturation of sperm in a man.

2 How would you begin to investigate experimentally the conditions which (i) induce flowering in a species of plant and (ii) promote reproductive behaviour in a mammal?

3 Comment on the following statements:
(a) Castrated animals do not possess secondary sexual characters.
(b) The ovaries of a pregnant woman may be removed without loss of the embryo.
(c) Pollen-grains of a flowering plant are equivalent to mammalian sperm.

4 Speculate on the reasons for the sporophyte generation assuming pre-dominance in the evolution of flowering plants.

5 Explain the role of the pituitary in the control of mammalian reproduction.

6 Compare the production of ovules with that of pollen-grains in a flowering plant.

9 Development

The development of an organism is the progressive elaboration of its body; this is usually accompanied by *growth* or irreversible size-increase. Growth and development are closely linked but, for convenience, development will be considered in this chapter and growth in the next.

Multicellular organisms begin life as one-celled zygotes which undergo repeated cell-divisions. The resulting cell products gradually diversify and differentiate (p. 74). Similar cells aggregate into *tissues* and different tissues into *organs* (e.g. heart and liver of animals, roots and leaves of plants) which compose the organism. The series of changes which gives rise to the organism's shape is called *morphogenesis*.

Animal zygotes develop into a hollow ball of cells which migrate as they multiply and form tissues which bulge and fold in specific ways to give the embryo its shape. In most animals, 'development' and 'embryology' are virtually synonymous, since the fundamental body plan is laid down in the embryo and, after birth or hatching, this body enlarges with very little developmental change. Exceptions are those animals such as tadpoles and caterpillars which undergo profound changes of form, called *metamorphosis*, between hatching and maturity.

The fully-developed plant embryo is not a whole plan for the mature body. Once a plant-cell has fully grown, it becomes encased in a fairly rigid cellulose cell-wall which prohibits further growth and development and the easy cell movement characteristic of animal cells. Post-embryonic plant growth and development depends upon *meristematic tissue*, or *meristems*, whose cells remain undifferentiated and capable of division. Some cells proliferated from such meristems differentiate and fulfil particular rôles, while others remain to divide further.

PLANT DEVELOPMENT

The course of plant development will be followed in this section from the time of fertilisation (see p. 115).

The secondary endosperm nucleus undergoes rapid mitotic divisions to produce many free nuclei within the embryo-sac. The nuclei gradually become enclosed within cell-walls, and these *endosperm cells* fill with stored food (mostly starch) supplied by the parent sporophyte. The embryo will use up this store of food in the endosperm when it begins to develop into a seedling. (In some seeds, the majority of the stored food is transferred to the *seed-leaves* or *cotyledons*.)

Embryo development

After fertilisation, the synergids and antipodals disintegrate. The zygote undergoes a series of mitotic divisions and forms a row of cells. The cell furthest from
the micropyle divides to form eight cells from which most of the embryo
develops. The other cells constitute the *suspensor*, growth of which pushes the
embryo into the endosperm (Figure 66). The embryonic cells continue to divide,
producing a ball of tissue which then develops two lobes. These lobes will become
the cotyledons. Cell-division and cell-enlargement elongate the basal region of
the embryo into the *hypocotyl*, below which will form the *radicle* (embryo root).
Above the cotyledons forms the *plumule* (embryo shoot), the base of which is
called the *epicotyl*. Cellular differentiation produces strands of long cells, called
procambium, within the tissues. Procambium cells will develop later into the first
xylem and *phloem* tissues which transport water and soluble nutrients respectively;
these are composed of specialised long, narrow cells.

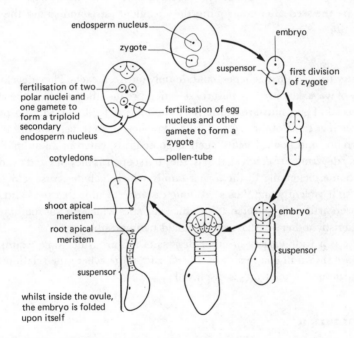

Figure 66   The development of a plant embryo
The embryo completes its development within the integuments of the ovule which
become the testa of the resulting seed

Seeds

While the embryo and endosperm develop and enlarge, the surrounding tissues
mature to produce a *seed*. The integuments fuse together and thicken into a *testa*
(seed-coat) which is continuous except at the tiny micropyle. The seed dries out
as it matures, and in this dormant, desiccated state it is able to remain alive under
adverse conditions. The breaking of dormancy is called *germination*.

Fruits

In addition to setting in motion seed development, fertilisation initiates a series of changes in the gynaecium which convert it into a *fruit*. The ovary-wall or *pericarp*, becomes succulent or dry and contributes to the dispersal of the contained seeds. Fruits may be classified as follows.

Succulent

Part of the pericarp becomes woody to form a 'stone' around the seed while the rest becomes succulent in *drupes* such as the plum and cherry. *Berries*, such as oranges, marrows and tomatoes are botanically distinct because the seed is surrounded only by its own hardened testa. There are 'false fruits' in which parts other than the gynaecium are incorporated into the fleshy structure. (Botanically, a fruit is only the mature gynaecium and the seeds which it contains.) Thus the strawberry is a fleshy receptacle, studded with *achenes* (see below).

Succulent fruits are attractive to birds and mammals, who reject or eliminate the seed after consuming the succulent parts and so aid the seed's dispersal.

Dry

Some dry fruits, called achenes, contain only one seed each. They develop in a variety of ways, favoured by natural selection because they enhance the chance of dispersal. Thus, nuts are attractive to animals as food, winged and plumed fruits, such as those of the ash and the willow-herb, can travel long distances in air-currents, and hooked fruits, such as goose-grass, catch in an animal's fur.

Many dry fruits are *dehiscent*, that is, they burst open as they dry and expose the seeds. Some, belonging to the *legume* family which includes gorse, clover and beans, split violently and flick seeds long distances. Cultivated peas and beans have been artificially selected to produce varieties which do not maintain this characteristic, otherwise harvesting would be impossible.

There is a wide variety of other designs of mature fruit, some examples of which are shown in chapter 17 of *The diversity of life*, where the relationship of floral structure to the fruits is displayed.

Germination

Once the seed ripens and is shed, the seed usually becomes dormant for a while. During *dormancy*, even those conditions normally favourable to germination will not make the seed germinate. Dormancy may be broken by various environmental conditions according to the species of seed involved, for example, cool, moist aerobic conditions, frost or passage through an animal's digestive system. Afterwards, the seed will germinate provided that it has water, oxygen and the correct conditions of temperature and light.

Germination begins when the seed rapidly takes up water. Initially, this imbibing of water occurs simply because large molecules in the testa are colloidal and naturally tend to absorb water. As substances inside the seed dissolve, more water is absorbed by osmosis. Enzymes in the seed are activated and begin to digest food-stores into soluble products such as sugars. These are translocated to the growing points of the embryo, giving them the energy necessary for growth and the materials with which to construct new cells.

The first visible sign of germination is the rupturing of the testa and the appearance of the radicle which grows downwards. Shortly afterwards, the plumule appears and pushes up.

Flowering-plants may be divided into the *monocotyledons* (or *monocots*), with only one cotyledon in the embryo, and *dicotyledons* (*dicots*) which have two cotyledons. Narrow-leaved plants such as the grasses are monocots, while broad-leaved plants are dicots. The tip of a monocot plumule is protected by a sheath called the *coleoptile* as it progresses through the soil. Once above ground, the plumule bursts through the coleoptile. Dicot plumules grow to the soil-surface curved in a hook so that the bent stem, rather than the delicate plumule, pushes through the soil first.

In *hypogeal* germination, the epicotyl elongates, carrying the plumule above the ground while the cotyledons remain below it. The food-store in the cotyledons nourishes the embryo until the plumule develops photosynthetic tissue. Broad-beans, oaks and maize undergo hypogeal germination. The hypocotyl elongates in *epigeal* germination so that the cotyledons, as well as the plumule, thrust above the ground. The cotyledons develop chlorophyll and become the first leaves to photosynthesise, the embryo being nourished until then by food from the endosperm. Beech, sunflower and marrow seeds exhibit epigeal germination (Figure 67).

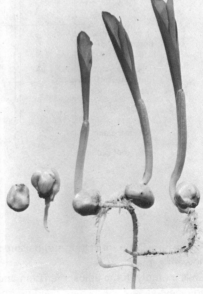

Figure 67   Epigeal germination of French bean and hypogeal germination of maize

Seedling growth

The tip of each young shoot and root contains an *apical meristem*, whose cells divide rapidly following mitoses of the nuclei. The cells which have divided most recently are closest to the meristem in the *zone of cell-division*. As cell-multiplication proceeds, the root- or shoot-tip advances and those cells which are 'left behind' enlarge in the *zone of elongation*. Water is absorbed osmotically by these cells, giving rise to small vacuoles full of sap. These vacuoles eventually coalesce to form a single vacuole in each cell. The cellulose cell-wall is thin and elastic and stretches as water accumulates so that the cell expands. The final shape depends on the way its cell-wall develops and on its neighbouring cells. Further from the tip than the expanding cells is the *zone of differentiation*. Here, the cells develop the characteristics typical of the tissues to which they will belong. The protection of the shoot-tip has already been mentioned. A root-cap of large, mucilaginous cells protects the root-meristem as it thrusts through the soil (Figures 68 and 70).

Growth from these primary meristems is called *primary growth* and results chiefly in an increase in length.

Shoot development

The shoot-tip is more complex than the root-tip, having whorls of *leaf-primordia* (bumps which will develop into leaves) surrounding the meristem. The tip

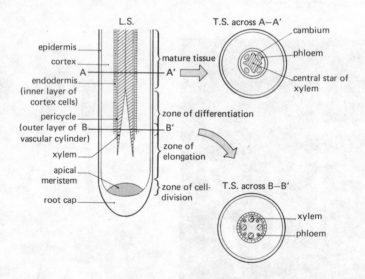

Figure 68   Diagram of longitudinal and transverse sections through a root-tip of a flowering-plant

Cells multiply in the apical meristem and then expand and differentiate as the tip advances. There is no sharp dividing line between elongating and differentiating cells. The transverse sections reveal phloem which cannot be seen in the plane of the longitudinal section

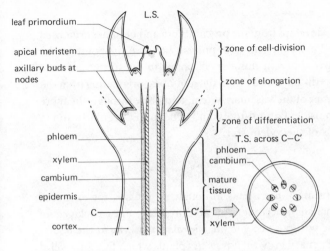

leaf primordium

apical meristem

axillary buds at nodes

L.S.

zone of cell-division

zone of elongation

zone of differentiation

phloem

xylem

cambium

epidermis

C

cortex

C'

T.S. across C—C'

phloem

cambium

mature tissue

xylem

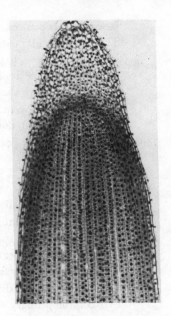

Figure 69 (above)   Diagram of longitudinal and transverse sections through a shoot-tip of a dicot flowering-plant

Figure 70 (right)   Section through the root-tip of an onion

consists of closely-packed *nodes* (regions at which leaf-primordia occur) with internodes between (Figure 69).

As the apical meristem progresses upwards, the developing leaves become more widely spaced. In the angle between the leaf and the stem is an *axillary bud* with the potential to develop into a branch stem. Leaf-growth occurs as the result of the activity of meristems along the leaf-margins. Since these cell-divisions are almost entirely in one plane, the leaf acquires its flattened shape.

Vascular tissue differentiates in the maturing stem. In monocots, scattered *vascular bundles* appear, each containing xylem and phloem, the phloem being nearer the outside of the stem. Vascular bundles in the stems of dicots are arranged in a circle, with the phloem on the outside. Eventually, developing leaf and stem vascular tissues connect up. Sugars are transported from photosynthesising leaves to the expanding stem-apex above.

Flowers

Under the appropriate conditions, including the correct photoperiod, flowering may be induced by bringing about a complete change in the shoot-meristem (see p. 109). Research suggests that there may be flowering hormones (called *florigens*) synthesised in the leaves under the 'inducing' conditions and translocated to the vegetative shoot-meristem which they transform. So far, however, no 'florigen' has been isolated. Once the transformation is accomplished, primordia of sepals, petals, stamens and carpels form in rapid succession at the shoot-apex instead of leaves. It is still not clear whether or not the change from a vegetative to a flowering apex is the result of the operation of a set of 'floral' genes which are not operative in the vegetative apex.

Root-development

The root's vascular tissue develops from the procambium and becomes organised into a central star-shaped axis of xylem with phloem between the points of the star. Different species have different numbers of points to the star of xylem.

Lateral-roots grow some distance away from the root-apex, originating from the *pericycle*, the outermost layer of the vascular system, and grow through the more superficial tissues to the outside. Lateral-roots tend to be spaced regularly and to occur opposite the points of the xylem star.

Secondary growth

Monocots possess only primary meristems although, in grasses, for example, portions of the apical meristem become separated from the apex by layers of maturing tissue and these portions become the *intercalary meristems* at the nodes.

In addition to apical primary growth, *perennials* (plants which live for several seasons) undergo *secondary growth* which increases their girth. This occurs in the root and stem by the division of cells in meristematic tissue, the *cambium*, situated between the primary phloem and xylem. Radial division of the cambium cells extends the meristem to form a cylinder separating xylem (inside) from phloem (outside). In each vascular bundle, cambium cells then divide tangentially, budding off *secondary phloem* on the outside and *secondary xylem* inside. Between the vascular bundles, *secondary parenchyma* cells bud off from the cambium; these constitute *medullary rays*, whose cells conduct materials radially through the stem. Additional medullary rays may form within the xylem of the stem.

Division of the cambium cells proceeds more rapidly on the inside of the cylinder than on the outside, resulting in more xylem than phloem cells. The phloem and the cambial ring push outwards, the cambium cells dividing radially so that the ring enlarges in circumference. The primary phloem is soon crushed as secondary phloem replaces it.

Cell-division in the vascular cambium ceases during winter and resumes in spring. The new growth produces large, thin-walled xylem vessels while narrower, thick-walled ones are produced later in the year. The summerwood is therefore denser than the springwood. A transverse section of a tree-trunk reveals concentric *growth rings* in the secondary xylem (*or wood*), each representing the end of a seasonal increment of growth. Growth-rings are sometimes called *annual rings* and are counted to estimate the age of the tree, but any check to growth, such as severe pruning, may produce an additional ring. At the centre of the trunk, the xylem becomes solid and ceases to conduct water but continues to support the tree (Figure 71).

As the vascular cambium becomes active, a more peripheral lateral meristem, the *cork-cambium*, appears in some plants. Division of these cells produces a layer of cork tissue on the outside and a spongy tissue called *phelloderm* inside. Cork cells become impregnated with fatty material which makes them waterproof. These layers confer protection and are continuously renewed as the stem expands; together with the phloem, they constitute the bark. The two lateral meristems enable the plant to grow in diameter (a process which disrupts the original

Figure 71   T.S. sections of tree trunks showing growth rings

epidermis or skin) without the flow of nutrients and water being interrupted.

Roots of certain herbaceous (non-woody) plants may be entirely without secondary growth and only the larger roots of trees undergo the process.

ANIMAL DEVELOPMENT

The zygote which results from fertilisation is a single cell with a large amount of cytoplasm. Repeated cell-divisions convert the zygote into a group of smaller cells in a process called *cleavage*. These cells are then arranged in layers during *gastrulation*, which is followed by a period of *organ formation*.

Cleavage

The pattern of cleavage is governed by the amount of yolk present. In sea-urchin eggs, with little, evenly distributed yolk, cleavage results in a collection of similar sized cells or *blastomeres* which arrange themselves into a hollow ball, called a *blastula*, surrounding a fluid-filled cavity or *blastocoel*.

Frog-zygotes, with substantial amounts of yolk, cleave unevenly. The first division, from animal to vegetal pole, bisects the grey crescent. The second division, also from pole to pole, is perpendicular to the first. The cleavage-furrows begin at the animal pole and cut swiftly through the cytoplasm until they are slowed down by the yolk in the vegetal regions. The third, horizontal, cleavage plane forms considerably above the equator so that four smaller cells are divided from four larger, yolkier ones below. From then on, divisions proceed more rapidly in the pigmented 'animal' half than in the 'vegetal' half. The animal half eventually contains more, smaller cells than the vegetal half (Figure 72(a)).

Bird- and fish-zygotes contain so much yolk that cleavage is restricted to a small cap of cytoplasm which collects around the animal pole after fertilisation. Repeated mitotic divisions produce a layer of cells, the *blastodisc*, which is equivalent to the blastula of other embryos. The blastocoel forms as a space between two layers (*epiblast* and *hypoblast*) of cells (Figure 72(b)). In birds, these early stages in development occur before the egg is laid, so they are difficult to study.

Mammalian zygotes cleave to form a cluster of cells called a *morula* which rearranges into an inner cell mass enclosed within a hollow *trophoblast* (Figure 72(c)).

During cleavage, the embryo does not increase in overall size, although a lot of DNA is synthesised, at the expense of cytoplasmic materials, to furnish each cell with a set of chromosomes.

Gastrulation

Following cleavage, cells move into new arrangements in a process called *gastrulation*, resulting in a three-layered embryo.

Sea-urchin

Gastrulation is simple to understand in sea-urchin embryos, unhampered by yolk. Development has been followed using time-lapse cinematography techniques. Cells at the vegetal pole begin to push into the blastocoel, creating a dimple which gradually elongates towards the animal pole. (The resulting structure may be envisaged as a soft ball which has been dented by thumb-pressure.) The new cavity so formed is the *archenteron* and it develops ultimately into the embryo's gut; it opens to the outside through the *blastopore*. The cells which now line the archenteron are called *endoderm*, while those forming the outer layer of the embryo are *ectoderm*; these are two of the three fundamental *germ-layers* from which all the adult's tissues will be derived. The archenteron cells move inwards together and the innermost ones reach out with long filamentous processes towards the inner surface of the ectoderm. These processes 'feel about' until they collect near the animal pole, when they contract, drawing the archenteron in (Figure 73).

As the archenteron begins to form, cells which will give rise to the middle germ-layer, the *mesoderm*, begin to migrate singly into the blastocoel from the advancing archenteron tip. The resulting embryo, or *gastrula*, transforms rapidly

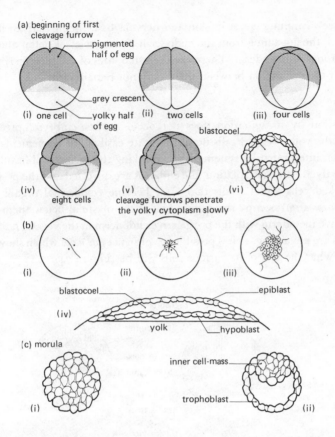

Figure 72(a) Stages in the cleavage of a frog-egg
(i) and (ii) are viewed from the middle of the grey crescent and (iii), (iv) and
(v) from the side of the crescent. (vi) is a sagittal section through the blastula.
(b) Cleavage of a bird's egg
The diagrams show only the nucleated cytoplasm on top of the yolk. (i), (ii) and
(iii) are viewed from above; (iv) is a sagittal section.
(c) Section through a mammalian morula at the end of the cleavage, (i), and after
reorganisation into an inner cell-mass and trophoblast, (ii)

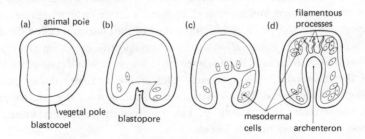

Figure 73 Gastrulation in the sea-urchin embryo
Mesodermal cells migrate into the blastocoel while the archenteron appears as an
invagination of tissue at the vegetal pole

into a free-swimming larva, its skin and nervous tissue being derived from the ectoderm, the gut-lining from the endoderm and muscles, skeleton and blood-system from the mesoderm. These are the usual fates of the three germ-layers, although the distinction between them may not remain clear-cut.

Frog

Gastrulation by invagination does not occur in frog-embryos, presumably because the yolky vegetal cells do not migrate easily. The movements of cells have been interpreted by systematically marking the surface of blastulae with harmless dyes. After gastrulation, the embryos are dissected and the positions of the marked cells found. Cells that later become ectodermal organs (called *prospective ectoderm*) occupy the animal half of the blastula; below them lies the prospective mesoderm with the prospective endoderm in the vegetal half. These positions are unvarying, so it is possible to construct *fate maps* which show on the blastula what the fate of its various cells will be (Figure 74).

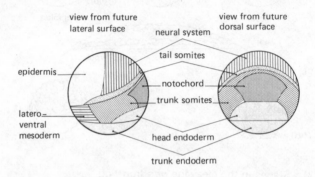

Figure 74   Fate map of an amphibian blastula
The somites are mesodermal, the epidermis and neural tissue ectodermal

Gastrulation begins with the appearance of a curved groove in the yolky cells of the *dorsal* surface (that is, the surface which will form the back), close to their junction with the grey crescent. The groove is formed by the infolding of prospective endoderm. Gradually the groove deepens and extends *ventrally* (that is, towards the future under-surface) to form a circle, which represents the blastopore. The active point at which the inward shift begins is called the *dorsal lip of the blastopore*. As gastrulation proceeds, prospective ectoderm cells spread downwards from the animal half to cover the yolky vegetal-half cells while, at the lips of the blastopore, cells approach, tuck inside and spread out into the blastocoel. At the dorsal lip, a cavity forms, leading from the outside to the groove where material is tucking in. This becomes the archenteron and, as it enlarges, it obliterates the blastocoel.

By the time the lips of the blastopore form a complete ring, the blastopore has migrated towards the vegetal pole. Yolky endoderm fills the blastopore as a *yolk plug* which gradually disappears inside as the lips of the blastopore contract (Figure 75).

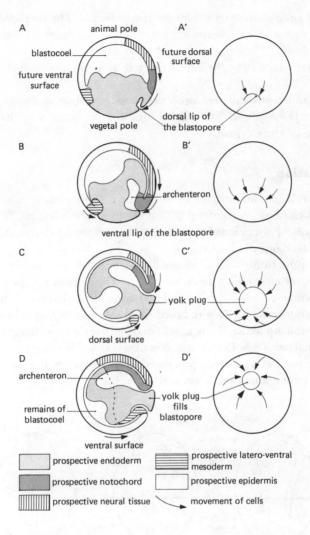

Figure 75 Gastrulation in the frog embryo

A, B, C and D are diagrammatic sagittal sections while A', B', C' and D' are surface views. The arrows indicate the direction of movement of tissues; the embryo also rotates so that its future dorsal and ventral surfaces come to occupy their ultimate positions.

Reference to Figure 74 reveals somite mesoderm which cannot be shown in sagittal section. This tissue tucks into the sides of the blastopore as a sheet whose limits at stage D are marked by the dotted line

Bird

When a hen's egg is laid, its blastodisc is in two layers, separated by the blastocoel, which does not extend to the edge of the disc. Seen from above, the zone with the blastocoel beneath is semi-transparent and called the *area pellucida*; it is surrounded by a ring of blastodisc tissue resting directly on the yolk: the *area opaca*. The embryo arises only from the area pellucida. The area opaca develops into the *yolk-sac*, which surrounds the yolk and mediates its transfer to the embryo.

Gastrulation occurs by the inrolling of cells along a groove called the *primitive streak*.

Three apparently different processes produce similar end-products: three-layered embryos whose tissues are distributed in such a way that they can interact to produce organs.

Neurulation

After gastrulation, an intricate series of morphogenetic movements is set into action which results in the development of the embryo's organs. There are too many details for discussion of them all, but the first, the development of the *neural tube*, is of fundamental importance.

Neural-tube tissue, which eventually becomes the central nervous-system, arises from the ectoderm of the dorsal surface. The diagrams (Figure 76) refer to the frog embryo; other species undergo equivalent changes. Ectoderm first condenses into a *neural plate* with raised edges called *neural folds*. The entire plate gradually rolls up so that the neural folds come together and fuse, resulting in a longitudinal *neural tube*. Ectodermal skin tissue extends over the tube which sinks below the surface. The whole process is called *neurulation* and the neural tube later differentiates into the brain and spinal cord.

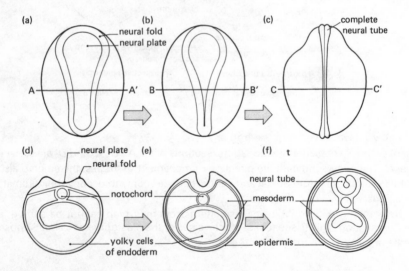

Figure 76 Neurulation in the frog
The upper row of diagrams shows dorsal views of the whole embryo at three different stages, while the diagrams below represent the corresponding vertical sections taken along lines A–A', B–B' and C–C'

Organ-formation

Most organ-formation results from differentiation of the mesoderm. The ectodermal neural tube gives rise to the brain; the peripheral nerves and the epidermis (outer layer of skin) are also of ectodermal origin. The gut-lining is endodermal.

During their development, most animals develop a fluid-filled cavity, the *coelom*, within the mesoderm, part of which becomes the adult's body-cavity.

Blocks of mesoderm, called *somites*, on either side of the notochord contribute to the *dermis* (inner layer) of the skin, to the backbone, which replaces the notochord, and to the skeletal muscles. In the somites, the coelom disappears.

Lateral plate mesoderm which surrounds the gut does not become segmented into blocks but becomes separated into two layers by the coelom. The inner layer surrounds the gut and also contributes to the formation of the heart. The outer layer forms part of the body wall and contributes to the limb muscles. Between the somites and the lateral plate mesoderm are small blocks of mesoderm which develop into the kidneys (Figure 77).

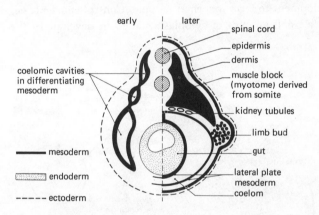

Figure 77   Diagrammatic T. S. through a chordate embryo showing early and later stages in the differentiation of the mesoderm

MECHANISMS OF DEVELOPMENT

Description of embryological events dates back over a hundred years, but it is only relatively recently that solutions have been sought to the problem of *how* the profound changes are brought about. We are still far from understanding the forces that govern changes of shape and form, but this is currently a rewarding area of research.

The translation of genetic instructions

In the course of development, groups of cells become restricted to some predictable pathway as they differentiate. In some species, even areas of the undivided, fertilised egg can be identified as those which will become the epidermis, say, or the notochord. In the early stages of development, it is possible to divert a cell from its normal course.

Nuclear transplants

Attempts have been made to find out what factors set a cell on its particular line of differentiation by transplanting its nucleus into a different sort of cytoplasm. Beginning in 1952, Briggs and King carried out a series of experiments in which they took nuclei from the cells of embryo frogs and injected them into enucleate uncleaved eggs (i.e. those having had their own nucleus removed). The ripe egg receiving the nucleus was taken from the oviduct and pricked with a glass needle to activate it. Its nucleus was removed. A cell from an embryo frog was then sucked into a fine pipette whose internal diameter was such that the cell-membrane broke, liberating the nucleus. The nucleus and a little cytoplasm were injected deeply into the enucleate egg and the hole was sealed (Figure 78). The majority of eggs so treated began cleaving and a small proportion completed their development to form a normal tadpole. Control enucleate eggs, injected with only a small amount of cytoplasm, failed to develop. In the successful

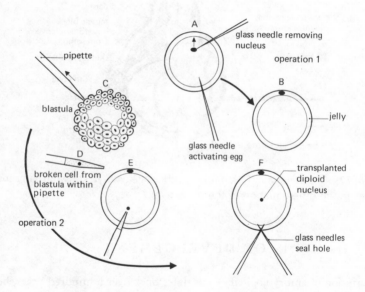

Figure 78 Nuclear transplantation in frog embryos
Operation 1 enucleates an unfertilised egg by pushing the nucleus into the surrounding jelly. A second needle pricks the cytoplasm, so activating the egg. Operation 2 removes a cell from the surface of a blastula with a fine pipette. The cell ruptures within the pipette, liberating the nucleus (D); the nucleus is injected into the enucleate egg (E) and the leakage through the remaining hole is cut off with a pair of glass needles (F)

transplants, the greater part of the cytoplasm involved came from the uncleaved egg, while the nucleus was one which, left in its original embryo, would have given rise to a differentiated group of cells with only limited specific functions (e.g. the gut lining). Transplanted, that nucleus was capable of multiplying many more times into daughter nuclei which could differentiate into all the kinds of cells required in the adult. The conclusion is that, during differentiation, cells do not lose any genetic material but each cell retains all the information necessary to create a complete new organism.

Cytoplasmic differences in eggs

If cells do not differentiate because they lose portions of their genetic material, it seems reasonable to look to the cytoplasm for the reasons why some parts of an embryo develop into one organ and not another.

Some eggs show regional differences in cytoplasm before they are even fertilised: the frog's egg is an example. Such eggs may be halved experimentally and each half, once fertilised, can develop into a normal embryo. There is one proviso: when the egg is halved, it must be divided, as in first cleavage, so that each half receives a share of grey crescent. Division in such a way that the grey crescent is all in one half results in that section only developing normally. Clearly, the cytoplasm is regionally differentiated from the outset.

In the usual course of development, cleavage subdivides the cytoplasm so that nuclei are surrounded by different regions of it. Once this first difference is established between the cytoplasmic environments of various nuclei, the possibility exists of activating and inhibiting different genes in different cells. Those genes which are active then can synthesise new substances which result in selective activation or inhibition of different genes. A series of reciprocal interactions between the genome and the changing cytoplasm sets the cell on a path of differentiation quite unlike that of a cell from a different part of the same embryo. Nuclear transplants work when a nucleus, with all its genes, and hence all its potentialities for development, still intact, is placed in a different cytoplasmic environment to which it is capable of responding. Nuclei from young embryos (blastulae or early gastrulae) more frequently allow complete development of the egg to which they are transplanted than do the nuclei of older embryos. The genes of the older embryos have been so long 'switched off' that it is more difficult to re-activate them.

Induction

Whether or not a tissue follows a particular line of development often depends upon its relationship with surrounding tissues. Experiments have been performed on newt blastulae of two differently pigmented species in which regions of prospective epidermis and prospective neural tube were exchanged. The fate of the transplanted cells could be followed because of their difference in colour. In each recipient blastula, the cells developed into the same tissue as their surroundings, that is, transplanted prospective epidermis developed as neural tube and prospective neural tube developed as epidermis.

If the same experiment is performed only two days later, when gastrulation is complete, the transplants retain the character they would normally have assumed without the operation; prospective neural tube develops as neural tube although it is moved to 'epidermis' position, and prospective epidermis develops as epidermis. Obviously, some important change accompanies gastrulation.

Further investigations show that, if prospective neural tube cells are prevented from coming to rest on top of the mesodermal archenteron roof after gastrulation, the neural tube fails to develop. Moreover, if prospective epidermis is placed over this mesoderm, it will become neural tube. The mesoderm is said to *induce* neural tube development. The dorsal lip of the blastopore gives rise to the mesoderm and has marked inducing properties in the early gastrula. If the dorsal lip is removed from one embryo and inserted into the blastocoel of another, it will develop in its new position into notochord and muscle. The ectoderm which overlies it is induced to become neural tube; sometimes a nearly complete additional embryo develops, attached to its 'host'. These discoveries illuminate the experiments with newt embryos: before the prospective neural tube is induced by the mesoderm, it will develop in a manner appropriate to its surroundings; afterwards it is *determined* and will become only neural tube tissue.

The induction of the neural tube by the underlying mesoderm is the first in a series of *embryonic inductions* in amphibian development. The dorsal lip of the blastopore from which this mesoderm develops is therefore called the *primary inductor* or *organiser* of the embryo. Once this induction has been performed, others may occur; the front end of the neural tube (now called the fore-brain), for example, develops an eye-cup which induces a lens to form in the ectoderm which overlies it. Surrounding tissues are then induced to form the transparent cornea and the tough eyeball coat. Such sequences are hierarchical; if one step is omitted, then subsequent ones do not occur.

Changes within cells

The changes in shape that tissues undergo as development proceeds are baffling in their complexity. Sheets of cells curve, invaginate and bulge outwards, others migrate long distances before settling down. The forces which generate these changes are poorly understood but, recently, efforts have been made to relate the alterations in form to changes in individual cells.

Many cells have the ability to contract – the cells of the sea-urchin archenteron which reach out long contractile filaments, have already been mentioned (p. 138). Neural tube formation also depends partly on local contractions of the upper surface of the cells in the centre of the neural plate, tending to curl the surface inwards. This is accompanied by an elongation of the cells in a direction perpendicular to the surface. Contractions within cells may be brought about by microfilaments with properties similar to those found in muscle.

The fact that sea-urchin archenteron cells stop moving when they contact the inner surface of the animal pole cells is also important. Cells make selective contacts with others, according to their mutual adhesiveness. Experiments have been performed in which separated embryonic cells, for example of heart and liver, are mixed together. In time, cells within the clump rearrange themselves so that all the heart cells are centrally placed, surrounded by liver cells. Each kind of cell has a particular relative affinity for cells of its own and other kinds; such affinities will guide the contacts which cells make and retain during development.

Polarity

Individual cells might be able to form specific recognitions but, if there were no way of knowing which was to be the head of the embryo and which the tail, that is, no knowledge of *polarity*, the result might still be chaotic.

Organisms obviously *do* possess a system of polarity and, what is more, individual cells may retain a 'knowledge' of their place in the system even after they have been separated from it. If the small pond animal, *Hydra*, is broken into its constituent cells and those cells heaped into a mass, the *Hydra* may reassemble. Its head, tentacles and foot form in the correct positions, just as before. Moreover, if specific cells are derived from different parts of a *Hydra*'s body and aggregated together, those which derived from the original head become the head of the regenerated animal, while those from the foot become foot. It appears that cells 'know' their relative positions because a gradient of some chemical emanates from the *Hydra*'s head end and the concentration to which a particular cell is exposed correlates with its position in the system. The 'head substance' causes the rejection of another head grafted close to it; the graft will only 'take' far from the original head.

Similar gradients within the developing organs of an embryo may help to establish the order in which the parts are laid down. Another possible mechanism exists. In the development of, say, an embryo chick's wing, differentiation spreads from the shoulder to wing-tip. In 1973, Lewis Wolpert suggested that cells are informed of their position in the limb-system by the time they spend (and hence the number of cell-divisions) in the growing, progressing tip. The cells which divide the fewest number of times differentiate as shoulder; those that divide the most, as wing-tip.

Recapitulation

All classes of vertebrates show remarkable similarities in their early embryological development; all possess, at some stage, a notochord, gill-pouches and a tail. This fact is evidence of their common evolutionary origin. Many of the 'ancestral' features are lost or converted to other functions before birth or hatching; mammals, for instance, develop the eustachian tubes (from the throat to the middle ear) from the first pair of gill-pouches. In 1868, the German biologist, Haeckel, made famous the *principle of recapitulation*: 'ontogeny recapitulates (or repeats) phylogeny'. This means that, during its embryological development (ontogeny), an animal retraces its evolutionary history (phylogeny). This theory is no longer taken entirely literally, but it is certainly true that embryos retain features of their presumed ancestors long after these features have been lost from the adult form. It seems likely that many 'primitive' structures provide inductive stimuli for the development of other organs – as the notochord induces the neural tube. Although the notochord is superseded by the vertebral column as an adult skeletal structure, it retains its inductive function, without which the nervous system would fail to develop.

EMBRYONIC MEMBRANES

Embryonic membranes of the bird

After gastrulation, the yolk of a frog embryo is inside the tissues, but that is not the case with birds: the bird gastrula perches atop a globe of yolk. In the chick, a network of blood-vessels develops from the inner part of the area opaca from the second day of incubation. This is now called the *area vasculosa*. Islands of blood-cells appear in the area vasculosa, and blood-vessel connections are made with the embryo. Gradually, the blastodisc tissue, with its developing blood-supply, engulfs the yolk and becomes the *yolk-sac*. As the yolk is absorbed into the blood and transferred to the growing embryo, the yolk-sac shrinks.

The body of the chick embryo becomes separated from the yolk-sac by the formation of the *head-* and *tail-fold* which tuck under the embryo and elevate it. Eventually, only a *yolk-stalk* connects the embryo to the yolk-sac. Folds of area pellucida tissue also grow upwards and eventually close above the embryo, so containing it in two *embryonic membranes*, the *amnion* and *chorion* (Figure 79). *Amniotic fluid* is secreted into the space enclosing the embryo, buoying it and protecting it from shocks and desiccation. The chorion continues to grow ventrally until the yolk-sac is completely enclosed by it. The fluid-filled space within the chorion is *extra-embryonic coelom* (i.e. coelom outside the embryo).

A third embryonic membrane, the *allantois*, originates as an outgrowth of the hind-gut. Small embryos excrete ammonia as their nitrogenous waste-product, but this is too toxic to produce in large quantities in the enclosed egg. As the embryo grows, it reverts to urea and then uric-acid as a waste-product; the insoluble uric-acid crystals are stored in the allantois until hatching. The allantois grows rapidly and penetrates into the fluid-filled space between the yolk-sac, the amnion and the chorion, its stalk being incorporated into the umbilical cord. By the middle of the incubation period, the allantois has spread underneath the entire chorion surface. Here it serves a second function: that of facilitating respiratory exchange. A network of blood-vessels develops on the outer surface of the allantois and the combined *allanto-chorion* comes to lie immediately below the shell membranes where it absorbs oxygen from the atmosphere, by diffusion through the porous membranes and shell, and excretes carbon-dioxide by the same path. The allantoic circulation continues until the chick hatches, when it ceases and the allantois dries up. Those structures (the yolk-sac, amnion, chorion, allantois and coelom) which do not form part of the embryo proper are called *extra-embryonic*.

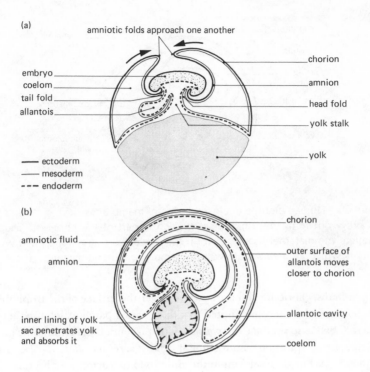

(a)

amniotic folds approach one another

chorion

embryo
coelom
tail fold
allantois

amnion

head fold

yolk stalk

yolk

—— ectoderm
—— mesoderm
--- endoderm

(b)

chorion

amniotic fluid

amnion

outer surface of
allantois moves
closer to chorion

inner lining of yolk
sac penetrates yolk
and absorbs it

allantoic cavity

coelom

Figure 79   The development of extra-embryonic membranes in the chicken
In (a) the amniotic folds are about to fuse to produce two separate membranes,
the amnion and the chorion, surrounding the embryo. In (b), the allantois has
grown and penetrated between the amnion and the chorion, largely filling the
coelomic space between them. The fused allanto-chorion spreads out to fit
beneath the egg-shell, which is not shown in the diagram

Mammalian membranes and the placenta

The development of the extra-embryonic structures varies slightly between
different groups of mammals.

The yolk-sac is so called because of its equivalence to the yolk-containing
structure in reptiles and birds, although in mammals it contains only fluid. It is
formed by the growth of endoderm and then mesoderm tissue around the inner
surface of the trophoblast. The yolk-sac is rudimentary in man.

In the rabbit, the amnion and chorion create folds which rise up to enclose the
embryo in an amniotic cavity as those of birds do. The human amniotic cavity
appears instead as a hollow in the mass of embryonic cells. The chorion also
forms the outer layer of the trophoblast which contacts the endometrium of the
uterus. Projections called *trophoblastic* (or *chorionic*) *villi* grow from its surface and
penetrate, branching, into the maternal tissues. The allantois, an expansion of
the hind-gut, grows out to line the chorionic villi and develops blood-vessels
serving the villi (Figure 80).

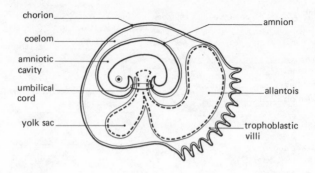

Figure 80   Human embryo enclosed in foetal membranes
Villi grow out from the trophoblast (which is part of the chorion). These villi penetrate the maternal tissues to form the placenta. The villi are lined with the allantois which connects them with the embryonic blood-system

In humans, chorionic villi are scattered over the surface of the trophoblast at first, but later they become limited to a disc-shaped patch which interacts with maternal tissues to produce the *placenta*. The placenta is the embryonic organ of attachment, respiration, nourishment and excretion and it also secretes hormones. Its blood-vessels run to the embryo in two arteries and return from it in a vein, all contained within the *umbilical cord*.

Placenta structure and function
The human blastocyst sinks into the endometrium of the uterus by the seventh day after conception. The uterine blood-vessels below the embryo become enlarged. As the embryo's trophoblastic villi develop, the superficial tissues of the uterus disintegrate beneath them, so that the chorionic villi dip directly into the spaces filled with maternal blood. Only three thin layers, the epithelium of the chorionic villi, their connective tissue and the lining of the embryonic blood-vessels, separate the mother's blood from that of the embryo. Transport of substances across the 'placental barrier' is therefore easier than it would be if the mother's blood-vessels and connective tissue remained intact. This form of placenta occurs in primates, including man, rodents, bats and some insectivores. In other groups of mammals, different numbers of layers of tissue remain intact between the mother's and the embryo's blood.

The placenta increases in size and vascularity, keeping pace with the embryo's growth, and the uterus expands accordingly. As pregnancy proceeds, the mother's blood leaving the placenta becomes progressively more de-oxygenated as the embryo's oxygen consumption increases. The thinness of the layers separating the embryo's and maternal blood and their large surface area (owing to the elaborate branching of the trophoblastic villi) facilitate diffusion of oxygen into the embryo's blood and carbon-dioxide away from it. In addition, the embryo's *haemoglobin* (the oxygen-absorbing blood-pigment) absorbs oxygen with greater avidity than does adult haemoglobin, and gives it up less readily.

Embryo haemoglobin is capable, therefore, of absorbing oxygen from its surroundings under conditions that make adult haemoglobin release oxygen.

Food-substances are conveyed from mother to embryo by diffusion and by *active* (i.e. energy-requiring) transport mechanisms. If the mother's own consumption of essential minerals, such as iron, calcium and phosphorus, is insufficient, her own body's reserves of these substances will be depleted as the embryo is preferentially supplied with them.

The antibody proteins, the gamma-globulins, which confer immunity to diseases, are also conveyed to the developing human *foetus* (an embryo older than two months, in humans). He is born with a 'passive immunity' to those diseases to which his mother is immune. This lasts for the first few weeks of his life, until he can develop an 'active' immunity of his own by exposure to the disease organisms. Some new-born animals, such as calves, receive their gamma-globulins as they suckle the first secretions of the mammary glands, called *colostrum*.

Certain drugs, nicotine from tobacco-smoking, and some viruses, such as that of German measles, can be conveyed to the embryo through the placenta, with damaging consequences.

Placental hormones

Gonadotrophin is produced from the chorion and the placenta also produces large quantities of oestrogen and progesterone. These hormones maintain the pregnancy and prepare the mother's body for birth and *lactation* (milk-production).

Birth

As the infant grows, its movements within the uterus become progressively restricted and it eventually settles against the mother's ring of pelvic bone. 'Labour', which precedes birth, begins with regular contractions of the uterus; these gradually increase in intensity and frequency. Hormones secreted by the infant, when ready to be born, cross the placental barrier into the mother's circulation to initiate this process. The uterus is also particularly sensitive to the hormone *oxytocin*, secreted by the mother's pituitary gland, at this stage. Oxytocin is a stimulant of involuntary muscles and it aids the delivery of the infant by increasing the uterine contractions. Compression bursts the bag of amniotic fluid and the baby is gradually squeezed into the outside world. The placenta detaches and follows after. Usually, breathing begins within seconds and glucose and fat reserves within the baby's body are mobilised as nourishment until suckling begins.

Circulatory adjustments at birth

Adult mammals possess a double circulation in which the left-hand side of the heart pumps blood around the body where its oxygen is given up to the tissues; this blood returns to the right side of the heart. The right-hand side pumps blood to the lungs where it absorbs oxygen; this blood returns to the left of the heart. In the foetus, the placenta, which is on the 'body' half of the circuit, is the site of oxygenation of the blood. Oxygenated blood therefore returns to the *right* side of the heart, but more than half of it is immediately diverted into the left atrium, instead of the right, via a hole called the *foramen ovale*. A flap acts as a valve which ensures that the blood passes only one way through the foramen ovale. The left atrium also receives a small quantity of blood returning from the non-functioning lungs. Blood from the left atrium is conveyed to the left ventricle and thence to the *aorta* (the main artery) and back to the placenta and the rest of the foetus' body. The right atrium receives some oxygenated blood from the placenta and some, which is de-oxygenated, from the foetus' body. In an adult, this blood would be pumped to the lungs, but the embryonic lungs have a high resistance to blood-flow and receive only a small blood-supply. The majority of the blood passes from the pulmonary (lung) artery through the *ductus arteriosus* into the aorta (Figure 81).

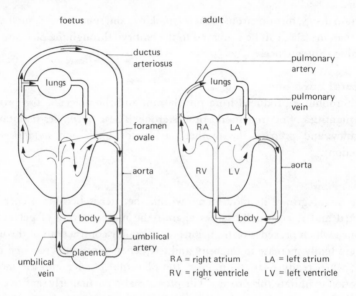

Figure 81   Foetal and adult circulation
In the foetus, most of the blood is diverted from the lungs by the ductus arteriosus and a hole, the foramen ovale, between the atria. Oxygen is supplied by the placenta which is on the 'body' circuit. When the umbilical cord is cut, and the lungs expand with the first breath, pressure rises in the left atrium, shutting the foramen ovale, and the ductus arteriosus closes.

After birth, the umbilical cord is severed. Before this happens, the umbilical arteries and vein contract so that there is little loss of blood. The fall in oxygen tension in the baby's blood and its decrease in temperature as it contacts the air for the first time, stimulate the respiratory centre in its brain and the infant makes its first inspiration; it takes a much greater-than-normal effort to fill the lungs for the first time. Blood-flow increases rapidly to the expanded lungs as they now have less resistance; more blood therefore returns to the left atrium from the pulmonary vein. At the same time, less blood returns to the right atrium since the placenta is now removed from the circuit. Greater blood-pressure in the left atrium than the right then causes a valve-like flap over the foramen ovale to close. Eventually, the flap grows over, so sealing the hole completely; if this does not happen, the baby has a 'hole in the heart', which may require surgery later in life. The relatively decreased resistance of the lung-circuit also causes the blood-flow to change direction in the ductus arteriosus which contracts in response to an increased oxygen-tension in the blood. Blood is now temporarily shunted towards the lungs instead of away from them. This may counteract the relative inefficiency of the newborn's lungs. Within a day or two of birth, the ductus arteriosus is blocked by the growth of fibrous tissue and the adult-type circulation is established.

METAMORPHOSIS

Some amphibians and insects undergo a striking change in form, called *metamorphosis*, during the course of their lifetime. The changes, under hormonal control, prepare the animal for a totally different lifestyle and require a major reorganisation of the organ-systems of the body.

In amphibians, metamorphosis is controlled by *thyroxine*, produced by the *thyroid gland*. Precocious metamorphosis can be induced by feeding tadpoles with additional thyroxine and removal of the thyroid gland prevents metamorphosis altogether. In common with virtually all other endocrine glands, the thyroid's production of hormone is regulated by the pituitary, which receives 'instructions' in the form of *neurohormones* secreted in the brain. Thus, external events, such as changes in the temperature and daylength, stimulate metamorphosis by leading to neurohormone secretion.

Insects will not metamorphose while their blood contains appreciable quantities of *juvenile hormone*, which is secreted by glands in the brain called *corpora allata*. As the larval insect ages, the corpora allata secrete less juvenile hormone. In the absence of juvenile hormone, the moulting hormone, *ecdysone*, produced by the prothoracic glands, induces the epidermis to secrete the adult cuticle.

Profound biochemical changes, affecting all the organism's cells, accompany metamorphosis. It appears that appropriate levels of the necessary hormones activate genes which synthesise enzymes required in the adult, whilst, presumably, repressing those concerned with forming larval structures.

The sequential activation of dormant genes to produce RNA was studied by Beerman and Clever. Following the injection of midge larvae with ecdysone, chromosomes in the salivary glands were observed to 'puff' or swell up at specific points whilst synthesising RNA at the site of the puffs (Figure 37 (a)). The mechanism which 'switches on' a set of adult genes at a particular point in an organism's life continues to be investigated.

QUESTIONS

1 Describe how you would investigate what factors induce the germination of a seed from a newly discovered species.
2 Explain how the frog embryo is shaped from the time of cleavage until neurulation.
3 How do mammalian extra-embryonic structures compare with those of birds?
4 Discuss the development of tissues within the elongating shoot of a dicotyledon.
5 To what extent do you consider the mechanisms concerned in development have been determined experimentally?

10 Growth and Regeneration

The growth of a living organism may be defined as its irreversible increase in size. Usually, such growth is accompanied by developmental changes including the differentiation of cells, but here size-increase alone will be considered.

'Size' is often difficult to measure, since living organisms have complex shapes. Various parameters may be chosen according to the material: length, width, area, volume, fresh- or dry-weight (that is, the weight of living tissue after heating at 100° C until a constant weight is reached) being the usual ones. Each of these parameters measures something different, and there is not necessarily a straightforward relationship between them. When a seed begins to germinate, for instance, it takes up water before synthesis of tissues begins, it has increased in weight and volume, but not 'grown' in the real sense. Afterwards, it increases in length whilst decreasing in dry-weight as stored food is metabolised. Only when the young seedling begins to photosynthesise does it increase in length, fresh- and dry weight simultaneously.

As the amount of cytoplasm increases in a growing organism, so does the protein-content of its tissues, and this can be measured with fairly complicated equipment. A simpler measurement is that of total nitrogen content, since about one-sixth of the weight of protein is nitrogen. The disadvantage of these criteria of growth (and one which applies to that of dry-weight measurements) is that they can be made only after the death of the organism concerned. The pattern of growth of a particular kind of organism can be charted only by analysing statistically the dry-weight or protein- or nitrogen-content of numerous similar organisms which were killed at different stages.

PATTERNS OF GROWTH

The time-course of the growth of three plants, measured in different ways, in shown in Figure 82. In each case, the *growth-curves* are S-shaped or *sigmoid*; that is, the size increases at first slowly, then rapidly and finally slows again to a halt. The overall growth is brought about by cell-multiplication and cell-growth, two processes which are intimately interrelated within the body of a multicellular organism.

Increases in cell numbers are easily investigated in populations of unicellular organisms such as the green alga *Chlorella*; these increases may be compared with those occurring within a multicellular body. Each cell is approximately spherical, grows to about 0.005 mm in diameter and then divides into two

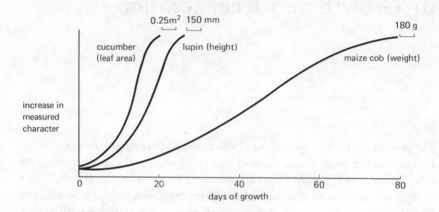

Figure 82   Growth-curves of leaf-area in cucumber, height in lupin and weight in maize cob
In each case, the curve is sigmoid

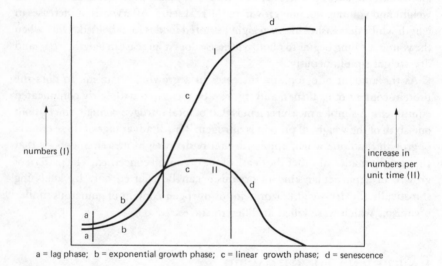

a = lag phase;   b = exponential growth phase;   c = linear growth phase;   d = senescence

Figure 83   Graph of population-increase in *Chlorella* culture (I) and a population growth-rate curve (II)

daughter cells which grow and divide in turn. Cells in the colony divide at different times so that the increase in total numbers is continuous. If the number of cells in a well-nourished colony is measured at fixed time-intervals (either by sampling and counting cells under a microscope or by measuring the density of the *Chlorella* culture), a graph of population-increase closely resembles the growth-curves of Figure 82. Figure 83 also shows the *growth-rate*, that is, the increase in number of cells per unit time, plotted against time. After a short lag-phase, (a), the increase in cell-numbers is *logarithmic* or *exponential*, (b), that is, its rate of growth increases continuously. Then follows a linear phase, (c), in which the rate of growth is constant and a phase of declining growth-rate, (d), during which the number of cells in the colony attains a maximum. Phase (d) is sometimes called *senescence* and results from the depletion of nutrients or the accumulation of toxic waste-products. If a few cells are removed from a senescent colony and sub-cultured in fresh growing medium they will increase in numbers as before.

The population-increase of a colony in its exponential-growth phase can be expressed simply in mathematical terms:

Let n_0 be the number of cells initially.
Let n_t be the number of cells after time t.
Let x be the number of generations during time t.

The number of cells in the colony after one cell-division will be $n_0 \times 2$.

after two divisions: $\qquad n_0 \times 2 \times 2 \qquad$ or $\qquad n_0 \times 2^2$

after five divisions: $\qquad n_0 \times 2 \times 2 \times 2 \times 2 \times 2 \qquad$ or $\qquad n_0 \times 2^5$

and after x divisions: $\qquad n_0 \times 2^x$

So $\qquad n_t = n_0 \times 2^x$

Therefore $\qquad \log n_t = \log n_0 + x \log 2$

log 2 is constant

$\therefore \qquad \log n_t = \log n_0 + \text{constant} \times x$

It is impossible to count the number of generations accurately, but easy to measure time-intervals, so

$$x = t/g$$

where g is the *generation-time*, or the time between the division of one cell and the division of its daughters.

So $\qquad \log n_t = \log n_0 + \text{constant} \times (t/g)$

g is a constant for a particular colony, so

$$\log n_t = \log n_0 + \left(\text{constant}_1 \times \frac{1}{\text{constant}_2} \right) \times t$$

The term in brackets is itself a constant, which may be called K.

i.e. $\qquad \log n_t = \log n_0 + Kt$

If the log of cell-numbers is plotted against time, a straight line will be obtained if the rate of population-growth is truly exponential (Figure 84). The log graph levels off as the colony approaches senescence.

While this mathematical model applies quite well to the exponential phase of size-increase of a cell-population or a multicellular organism, the model is not followed by most organisms for long. In multicellular organisms, cells eventually differentiate and so do not retain the permanent powers of division as *Chlorella* cells do. The number of cells which stop growing and dividing increases as the organism ages; in mature vascular plants, growth is restricted to the meristems whereas mammals cease growing altogether at a certain stage. The rates of growth of different organs is usually co-ordinated and the organism's ultimate overall size is regulated in various ways.

The growth-curves of an arthropod and a mammal are illustrated in Figure 85. Each of them broadly follows the sigmoid pattern, but there are phases of slower growth corresponding, in the mammal, to the period immediately after birth, at the start of weaning and at puberty. They are correlated with dietary and hormonal adjustments. An arthropod's growth in volume is necessarily spasmodic, since the animal can swell up only in the period following a shedding of the rigid outer skeleton (*ecdysis*) until its new skeleton hardens. Those animals, such as amphibians and certain insects, which undergo a profound change in form called *metamorphosis* temporarily lose weight at that time (see p. 153).

GROWTH OF INDIVIDUAL CELLS

While animal cells are bounded only by a delicate living plasma membrane, plant cells are encased in addition in relatively rigid walls made of specially organised cellulose microfibrils. This difference accounts for a fundamental contrast in the way the cells grow. Following mitosis, an animal cell-membrane becomes constricted between the two daughter nuclei and the cytoplasmic connection between them narrows and disappears. The daughter cells may then wander about freely and can change shape. The cells absorb materials by various processes, undergo much protein synthesis and manufacture more plasma membrane as the cell grows.

Plant cells do not move apart after cell-division. Instead, they retain contact via slender cytoplasmic connections called *plasmodesmata* which extend through *pits* in the cell-wall that separates the daughter cells. After cell-division, a plant cell grows mainly by absorbing water and some small molecules. Most of this solution accumulates in the central vacuole so that the living cytoplasm is restricted to a thin peripheral layer occupying less than ten per cent of the cell's total volume. As the cell grows, the original wall stretches and becomes thinner, and new wall material is laid down. In a newly divided cell, most of the cellulose microfibrils are oriented more-or-less perpendicularly to the line of future cell-elongation. When the cell swells as it absorbs water, the expansion most readily forces the microfibrils apart (rather than extending them), often tilting them in the process. This results in a longer cell (Figure 86). Since adjacent plant cells are usually in contact with one another on all sides via plasmodesmata, it is important that they elongate in harmony otherwise the connections would be broken.

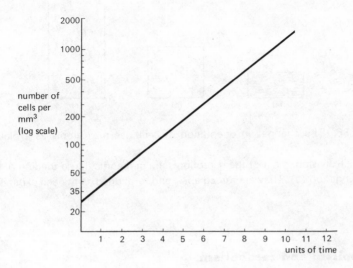

Figure 84   A straight-line graph obtained by plotting the log of cell numbers of *Chlorella* in the exponential growth-phase against time

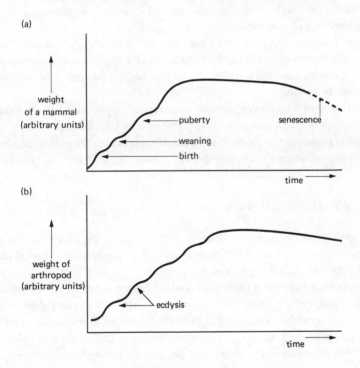

Figure 85   Growth-curve of a mammal (a) and an arthropod (b)
The checks in growth are caused by hormonal and dietary changes in the mammal and ecdysis in the arthropod

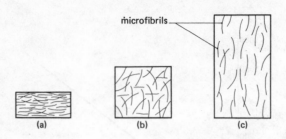

Figure 86   The changes in orientation of cellulose microfibrils as a plant cell enlarges
After cell-division, most of the microfibrils are perpendicular to the line of future cell-elongation (a). They are forced apart and re-aligned as the cell expands, (b) and (c)

Anabolism and catabolism

The syntheses of complex molecules from simpler ones are called *anabolic reactions*. Such syntheses require energy which is supplied by respiration in all tissues and by photosynthesis in suitably pigmented plant tissue in the light. The energy is trapped in the high-energy compound ATP, which is produced by both respiration and photosynthesis.

Three fundamental processes which require ATP-energy and occur during growth are the syntheses of nucleic-acids (which supply new chromosomes and RNA), proteins and carbohydrates (particularly those involved in cell-wall formation in plants).

The reactions which take place during respiration, breaking down foodstuffs to simple compounds and providing useful energy which is stored as ATP, are examples of *catabolic* reactions. While anabolism proceeds faster than catabolism, the cell grows; when the reverse is true, it dwindles and will eventually die.

PLANT HORMONES

Plant growth occurs in a co-ordinated fashion, indicating that some control is exerted within the plant. This internal control is affected in a general way by external factors such as light and temperature.

The contrast between a plant grown in full light and one grown in the dark is dramatic (Figure 87). The former is upright, sturdy, with expanded green leaves and short internodes. The latter is much taller, spindly, with underdeveloped leaves and a hooked apex; it is yellowish, owing to the absence of chlorophyll (which develops only in the light), and, because it cannot photosynthesise, it will die when the food-reserves from the seed are exhausted. Light has the paradoxical properties of inhibiting cell-expansion in the stem whilst promoting it in the leaves.

In order to affect a biological process, light must be absorbed by a pigment. If

Figure 87   Seedlings grown in the light and in the dark

recently germinated seedlings are tested with different wave-lengths of light, red
light of wave-length 660 nm is found to be most effective in causing the
development of characteristics normally found in seedlings grown in daylight.
Even very low intensities of light for short periods are sufficient. Investigations
have shown that the pigments involved are the phytochromes, previously
discussed in connection with flowering (p. 110).

Auxins

The sequence of reactions between the absorption of light and the promotion or
inhibition of growth is still not fully understood, but it involves *plant growth-
hormones*, chemicals synthesised by the plant, which elicit specific reactions even
in very low concentrations. One of the best-known hormones is *auxin*, which is
formed in the stem-tip and young tissues. Chemically, auxin has been identified
as *indole-acetic acid*, abbreviated to IAA. Related auxins with more restricted
ranges of function are also important to plant growth.

Charles Darwin was first to discover that the tip of an oat coleoptile perceives
the direction from which a light is shone on it, and that the expansion of cells
several millimetres below the tip is affected so that the coleoptile grows faster on
the shaded side and hence grows towards the light. He suspected that a growth
substance mediated between the tip and the growing cells; experiments in the
1920s and 30s by Went and Boysen-Jensen proved him correct. In a series of
delicate investigations, these research-workers demonstrated that the growth of a
coleoptile stopped if its tip were removed but, after resting the tip on a small
block of agar-jelly for a while, the agar could replace the tip in inducing the

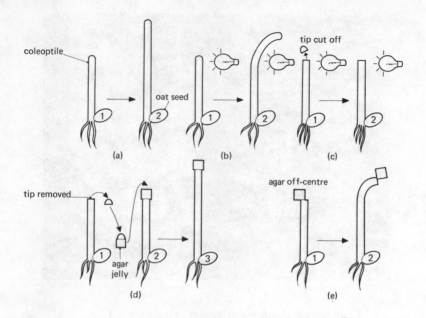

Figure 88   Investigations into the growth of oat coleoptiles
(a) shows the growth of coleoptiles in all-round illumination. In (b), the coleoptile bends towards light from one side. In (c), both growth and bending are eliminated by removing the tip. (d) shows the auxin being allowed to diffuse from the decapitated tip into an agar-block which then induces the stump to grow. A similar block, (e), placed off-centre causes greater growth on one side of the coleoptile than the other and hence bending

coleoptile to grow (Figure 88). Obviously a substance, later identified as auxin, had diffused from the coleoptile-tip into the agar-block. Putting the auxin-impregnated agar-block back on the decapitated coleoptile, off centre, resulted in the coleoptile growing faster on the side closer to the block. This response was so similar to the bending of a coleoptile towards light that it led to the hypothesis that light acts by partially destroying auxin. Now we know that light, absorbed by phytochrome, enhances the activity of an enzyme which destroys IAA. This explains the excessive elongation of seedlings in the dark and provides the mechanism whereby the shoots of plants position themselves in such a way that they receive maximum illumination.

Gibberellins

The gibberellins derive their name from the fungus *Gibberella fujikuroi* which attacks rice plants, causing them to grow excessively tall and fall over. Japanese biologists discovered that the disease-symptoms could be produced by treating healthy plants with cell-free extracts of the fungus. Obviously, some chemical was interfering with normal growth-processes.

Many gibberellins are now recognised, certain of which are active in particular species only. In general, gibberellins promote stem-elongation and, if applied to genetically dwarf strains of plants, they will induce growth comparable to that in normally tall varieties.

Cytokinins

In the 1950s, investigators who were culturing pith cells from tobacco plants discovered that although cell-division usually continued for only a short time in a normal culture-medium, it could be prolonged by the addition of coconut milk or yeast-extract. The chemical responsible for the activity was identified as a breakdown product of DNA and named *kinetin*. Other substances, both naturally occuring and artificial, which stimulate cell-division have been found, and they are collectively called *cytokinins* (*cytokinesis* is cell-division). High cytokinin-levels have been reported for many actively dividing tissues.

*

Natural plant growth seems to be regulated by an interplay of growth-promoting and growth-inhibiting hormones. (One of the best-known inhibitors, called *abscisic acid* or *dormin*, is potent enough to reduce the growth-rate of duckweed in concentrations of 1 part per 1000 million.) A change of environmental conditions may be the factor which sets off a sequence of hormone-syntheses. Research is proceeding in order to establish exactly how hormones exert their effects. In the cases of cytokinin and dormin, it is established that they accelerate and retard the rate of nucleic acid synthesis respectively; their control is exerted at the most fundamental level.

Limits to growth

There are limits to the maximum attainable size in both plants and animals. If an organism grows twice as high, broad and long as it was before, its surface area increases by 2^2 and its volume (and hence usually its weight) by 2^3. Transport systems convey materials from the outside to the organism's interior, remote from the surface, but they work efficiently only within certain distances. Supporting structures such as the leg-bones of animals and the trunks of trees must be disproportionately thick in large creatures, compared with small ones, in order to withstand their greater weight. This particular factor limits land animals to a relatively smaller maximum possible weight than land plants since animals must move about.

ANIMAL GROWTH HORMONES

As with plants, the ability of an animal to grow is basically dependent on its nutrition. Starvation will never produce such a large body as good feeding. In addition, as in plants, animal growth is controlled by *hormones*, chemicals produced in certain *ductless* or *endocrine* glands, distributed throughout the body in the blood and having specific functions elsewhere.

In mammals, the pituitary gland secretes a number of hormones which control the activity of other endocrine glands and also a growth-hormone, which has a direct effect on the growth of bones and soft tissues. Underproduction of growth-hormone results in a dwarfed animal whilst overproduction (sometimes caused by a pituitary tumour) produces gigantism. The *thyroid*, a gland next to the trachea in mammals, produces *thyroxin*, a hormone with powerful effects on the metabolic rate of all tissues. Deficiency of thyroxin during the normal growth period inhibits growth and sexual development and results in mental sub-normality. Other hormones also play their part in the growth process because of their influence on the metabolism of food.

The output of hormones from endocrine glands is maintained by feedback mechanisms. The secretion of *thyrotrophin* from the pituitary, for instance, induces the thyroid gland to increase its size and hence its rate of thyroxin-secretion. This in turn increases the body's metabolic rate, which tends to lower the output of thyrotrophin and so reduces thyroxin-production. A low rate of metabolism causes the pituitary to secrete more thyrotrophin, so stimulating the thyroid and increasing metabolic rate. Such a homeostatic mechanism maintains metabolism, and hence growth-rate, within narrow limits.

REGENERATION

Multicellular organisms outlive the lifespan of their constituent cells (with the exception of cells of the animal central nervous system) because there is constant replacement of cells that die. During growth, the production of new cells exceeds the death of old ones, and during senescence the reverse is true, but for most of a healthy organism's life cell production and loss are kept in balance. Certain tissues, such as mammalian skin and red blood-cells, have a rapid rate of turn-over while bone and cartilage cells are relatively long-lived.

Occasionally, the controls which operate to limit the number of cell-divisions in an organ are disrupted and cells continue to multiply, producing a *tumour* or *cancer*. Cancer cells are typically mobile and fail to aggregate into normal tissues.

Damage to a plant or animal may stimulate a renewed growth of tissue, called *regeneration*. Mammals have poor powers of regeneration, amounting to no more than wound-healing but some plants, such as the dandelion, can regenerate entirely from small sections of root. Indeed, research has shown that differentiated phloem cells may be removed from the root of a carrot and cultured in a growing medium, when single cells will multiply into colonies which differentiate into complete plants. Under normal conditions, damaged vascular plants regenerate tissues by renewed meristem activity. Similar phenomena account for the formation of new individuals without injury by vegetative reproduction.

Invertebrate animals generally have much greater regenerative powers than do vertebrates. The pond flatworm may be divided in two, and each half will regenerate another and make a whole flatworm. Lobsters and crabs frequently shed legs when trapped by them (or even when merely startled) in a process called *autotomy*. Later, a whole limb regrows from each stump.

Among vertebrates, the only adult animals capable of regenerating large appendages are lizards, which may shed their tail at the mere approach of a predator (the tail then wriggles, usually distracting the predator while the lizard escapes), and some amphibians. Amputation of a limb of a salamander is followed by regeneration of an epithelium over the cut surface. Within ten days the end of the stump enlarges into a *regeneration bud* containing many rapidly-dividing cells which appear like embryonic cells. Research has established that these cells derive from the neighbouring bone, muscle and connective tissue; they have *de-differentiated*. The bud grows out and the cells eventually re-differentiate into a limb-pattern similar to that laid down in embryological development.

The mystery remains as to why only some animals retain considerable regenerative powers. There would be obvious applications to medicine if methods could be found to stimulate similar activity in humans.

QUESTIONS

1 What experiments would you carry out to establish whether all the cells which contribute to the formation of a regenerated newt limb arise close to the cut surface?

2 The table below shows measurements of different characteristics of oat plants over a period of ten weeks from the time that the single-seeded fruits were sown.

| | | Growth in oat plants | | |
Days from sowing	Height (cm)	No. of leaves	Wet mass (g)	Dry mass (g)
0	–	–	0.0656	0.0400
7	–	–	0.1144	0.0290
14	9.54	1	0.2120	0.0331
21	17.0	2	0.3550	0.0440
28	25.3	3	0.8700	0.0952
35	38.0	4	1.9370	0.1770
42	48.5	8	4.0500	0.3800
49	55.0	10	8.3500	0.7600
56	65.7	13	17.0000	1.5000
63	78.5	26	30.3000	2.4300
70	85.0	48	60.8000	5.1000

(a) Draw a graph of measurements against time for each of the features shown.

(b) Comment in detail on the shape of each of the graphs you draw.

(c) State which of the graphs you regard as the best indicator of growth of the plants concerned, giving your reasons.

3 In what ways does the growth of plants differ from that of animals?

11 The Progression of Life

SPONTANEOUS GENERATION

For centuries the profound question 'Where did life come from?' had a very simple answer: life came from non-life. Recipés for bringing about the transformation were well known; maggots were made in old meat and mice in straw-filled coats. This theory that the animate arose by itself from the inanimate is called the theory of *spontaneous generation*.

Francesco Redi applied the new scientific method to the problem in 1668. He placed meat under a muslin cover and demonstrated that maggots did not develop when flies were denied access to lay their eggs. Spontaneous generation apparently did not work for large organisms. However, the invention of the microscope had revealed a seething world of microbes, and many scientists still argued that these arose from the non-living matter on which they lived.

Pasteur's experiment

The French Academy of Science offered a prize to anyone who could solve the spontaneous generation controversy. The challenge was accepted by Louis Pasteur.

It was easily demonstrated that both air and broth contained microbes and that, if the broth were boiled, the microbes were destroyed. The problem was to allow sterile broth to contact fresh air (as critics otherwise complained that microbes could not develop in its absence) without the air contaminating the broth with microbes from elsewhere. Pasteur's solution was elegant. He boiled broth in a flask whose long neck was drawn into a deep curve (Figure 89). Microbes already present in the flask were killed. As the flask cooled and contracted, air was pushed back in by the pressure of air outside. By the time the flask was cool enough to allow the incoming microbes to survive, the air moved so slowly that the slightly heavier-than-air organisms were deposited at the bottom of the bend. The broth remained clear for months, and microscopic examination of a sample failed to reveal spontaneously-generated microbes. Only when the swan-neck of the flask was broken did invading microbes turn the broth cloudy, so demonstrating that heating the broth had not ruined its ability to support life. Pasteur's work was repeated and verified, and in 1862 he won the French Academy's prize as the man who demolished the theory of spontaneous generation. Some of Pasteur's original swan-necked flasks can be seen at the Pasteur Institute in Paris; they are *still* sterile.

CHEMICAL EVOLUTION

The problem remained: if life does not arise spontaneously from the non-living, where does it come from? One hypothesis maintained that life has always existed and it arrived on earth as 'seeds' from elsewhere in the solar system. Another proposed that life was created supernaturally in the past. The third theory, put forward in the 1920s by the biochemists Alexander Oparin in Russia and J. Haldane in Britain, says that life arose on earth by a process of chemical and biochemical evolution. This is a sophisticated way of saying that life did, after all, arise from non-life. Unlike the pre-Pasteur spontaneous generationists, however, the modern theorists believe that life is no longer generated on earth naturally under the conditions which prevail. Many stages in the evolution of life as envisaged by Oparin and Haldane can be demonstrated experimentally, however.

A definition of what constitutes 'life' is not easy. Modern living forms obtain energy from the outside world for their growth, they are sensitive to stimuli and they reproduce. It is impossible to say at what point in chemical evolution an organised collection of molecules exhibited enough of these properties to be considered alive. It may be more useful to think of 'degrees of livingness' from the totally non-living to the fully alive, much as a chemist uses 'hydrogen-ion concentration' as an expression of the degree of acidity of a solution which may range from the very acid to the very alkaline.

Biochemistry has emphasised the remarkable unity of living matter. All organisms rely on twenty or so amino-acids from which they construct protein, and five bases, a phosphate and two sugars which make up nucleic acids. Life apparently depends upon the transfer of information between nucleic acid and protein. Moreover, the *metabolic*, or material-transforming processes which occur in living cells involve similar reactions in bacteria, plants and animals. This fundamental similarity is powerful evidence that only one line of chemical

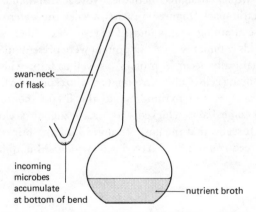

swan-neck of flask

incoming microbes accumulate at bottom of bend

nutrient broth

Figure 89   The simplest version of the Pasteur flask
The broth is sterilised by boiling. On cooling, the microbes are trapped in the bend and the broth remains uncontaminated

evolution gave rise to all living organisms – that 'life' arose only once. Much effort of research and speculation has gone into considering how it did so.

Stages in chemical evolution

Changes in the atmosphere

When the earth was formed, the large amount of hydrogen present combined with carbon to give methane (CH_4), with nitrogen to give ammonia (NH_3) and with oxygen to form water (H_2O). The primitive atmosphere was therefore non-oxygenic or *reducing*. Since today our atmosphere contains twenty-one per cent free oxygen, considerable changes have occurred. Initially, ultra-violet radiations from the sun dissociated water molecules into oxygen and hydrogen; the latter, being light, tended to escape into space. Much later, when living microscopic organisms had evolved and developed the art of *photosynthesis* (the process by which green plants convert solar energy into chemical energy) they emitted much more oxygen. Eventually, oxygen molecules in the upper atmosphere combined together to produce an ozone (O_3) (now called tri-oxygen) layer which acts as a filter to ultra-violet light. Oxygen accumulated beneath the ozone shield until it reached the equilibrium point which is maintained today.

Early synthesis

Many sources of energy were potentially available for the synthesis of molecules in the primitive, pre-biological earth. An important one was ultra-violet light, until that source was cut off by the development of the ozone shield. Others included heat energy and electricity in the form of lightning.

Miller's experiment

A classic experiment was performed by Stanley Miller in 1953 to investigate the rôle of electrical discharge as a source of energy in an atmosphere like that of the primitive earth. Miller circulated methane, hydrogen, ammonia and steam past a repeating electric spark (Figure 90). After a week the water was analysed for *organic* (carbon-containing) compounds. The results exceeded expectation. Four of the amino-acids commonly found in protein were present (along with several which do not naturally occur in protein) as well as fatty-acids and urea, also fundamental to living cells. Other investigators who repeated Miller's work have added sugars, purines and pyrimidines to the list of spontaneously-formed chemicals, and some have achieved synthesis using ultra-violet light as an energy-source. It seems that the fairly complex building-bricks of which living matter is composed could indeed have been synthesised in quantity by purely chemical means.

Primitive 'soup'

Assuming that organic compounds accumulated in water, endless chemical reactions would have occurred, synthesising and breaking down molecules. The complex organic mixture that resulted has been called a *soup*. 'Chemical selection' must have been in operation in the soup, i.e. those molecules survived

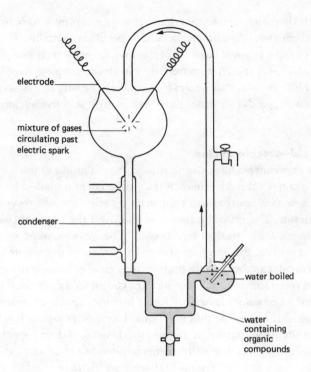

electrode

mixture of gases
circulating past
electric spark

condenser

water boiled

water
containing
organic
compounds

Figure 90   Diagram of the Miller experiment
Boiling water created steam which circulated a methane, hydrogen and ammonia
atmosphere past a repeating electric spark and then a condenser. After a week, the
water was rich in organic compounds

which were best able to grow at the expense of their surroundings.

The earth was formed about 4600 million years ago, and the first definite
evidence of living organisms comes from rocks 3200 million years old. There was,
therefore, a very long period in which chemical evolution, first to form an
organic soup and then to form living organisms from it, could have occurred.
Since living cells contain a high concentration of large polymers, a dilute soup
from which to create them would not do. Any soup which accumulated in small
pools might have evaporated, so increasing its concentration and hence the
chance of the small molecules reacting together to form polymers. Clay particles
might have been useful in bringing about polymerisation, since clay adsorbs
amino-acids and allows them to join together into protein-like chains.

In a process of pre-biological natural-selection, those molecules which first
developed stable structures would be those which persisted longest and so
became the most likely candidates for incorporation into a life-system. Chains of
nucleic-acid bases are especially capable of stability because chemical bonds
form readily between adenine and uracil and between cytosine and guanine.
Cross-linking tends to turn single chains into clover-leaf shapes, a design still
employed by transfer-RNA (Figure 33 p. 68); its compact shape helps to
preserve it from damage. For natural selection to have favoured stable clover-

leaf molecules in preference to other shapes, the survivors must have been able to reproduce themselves. Again, nucleic acid is peculiarly suitable. If RNA-type loops open out, the exposed bases react most readily with their complementary partners (A-U; C-G) chosen from free bases to give a complete complementary molecule. This, in turn, can generate a copy of the original molecule. Such ready, accurate reproduction must have been of great selective advantage in the soup.

Nucleic-acid–protein cycles

Laboratory experiments have demonstrated that solutions of nucleic-acid can proliferate in a test-tube. Moreover, if the environment is made adverse for the molecules, new versions that can cope arise by selection and reproduction of suitable mutants. The crucial question is: 'how did the essential co-operation between nucleic-acids and protein begin?' The answers must be based on informed guess-work. Proteins cannot reproduce their own structure in the way that nucleic-acids can, but certain proteins may have enzymatically speeded up nucleic-acid reproduction (such enzymes are known today). A point must have been reached when a nucleic-acid molecule became capable of aligning a series of amino-acids. It is possible that individual bases, or groups of bases, already possessed an affinity for particular amino-acids because of their respective shapes and chemical attractions. Logically, then, members of a chain of bases may have each brought into line the amino-acid with which they reacted. Attraction between the aligned amino-acids may then have proved the stronger force so that a protein chain was formed and released. The random production of proteins in this way would be favoured by natural selection only if the proteins had a means of enhancing their own production. By a series of unknown intermediate steps, it might have happened that a particular nucleic-acid sequence began to manufacture a protein which functioned as an enzyme to speed up the rate of that nucleic-acid's own reproduction. Such a system would have an improved chance of survival compared with those with more slowly-reproducing nucleic-acids. Eventually, the best-organised nucleic-acid–protein cycle might have proliferated at the expense of its rivals (which probably used different base– amino-acid 'dictionaries') and subsequently developed in the first living cells. Alternatively, cells arose from a number of different cycles and only later did natural selection 'choose' one system in preference to others. In any event, living organisms today use the same, much elaborated, genetic-code for translating information from nucleic-acids into proteins: they must have had a common origin.

Such a theory is highly speculative and seems to require a series of very-unlikely coincidences. Scientists investigating the origin of life maintain that, given the vast time-span available and the fact that certain chance events increase the likelihood of others happening, the unlikely becomes inevitable.

Coacervates

A vitally important step in the evolution of life was the transition from molecules replicating in solution to molecules replicating in cells. It must have been a considerable selective advantage to the successful nucleic-acid–protein cycles to be separated in some way from the rest of the primitive soup. In this way, the vital chemicals could be concentrated without contamination from toxic materials. Oparin has demonstrated that organic solutions can spontaneously form droplets called *coacervates*. These small aggregations collect together because of the attraction between polymers of opposite electrical charges. A fatty skin may surround the droplet and maintain it intact whilst allowing certain materials to pass in and out. Coacervates have the ability to concentrate small molecules within themselves and enzymatic reactions can take place inside. If coacervates did form in the early stages of the evolution of life, they presumably grew in size as they accumulated materials from the soup of molecules which surrounded them. At a certain size they would have become unstable and broken up into smaller units. Natural selection would favour those whose 'daughter' coacervates retained a full set of the essential molecules, and those whose boundaries were strengthened by the evolution of true membranes.

Energy sources
The question of how energy was initially absorbed and mobilised for biological synthesis is difficult to answer. Chemical energy for synthesis in modern organisms involves *organo-phosphate unions*. A source of energy is needed to form such unions and the energy released on breaking them is regulated by a series of enzymes. The most commonly-used organo-phosphate compound today is the high-energy compound, *adenosine triphosphate* (ATP). Even inorganic phosphate has some energy-transferring properties which are enhanced by association with organic compounds. ATP could have evolved gradually from such an association.

Left-handed and right-handed molecules
Certain molecules occur in two chemically identical forms or *isomers* of one another which, in some cases, are left-handed and right-handed mirror images. Louis Pasteur made original observations on the left- and right-handed crystals of sodium ammonium tartrate, laboriously separating them by hand with the aid of a microscope. Separate solutions made of the two isomers behave differently in a beam of *polarised light*. (Light waves normally vibrate in every plane. When they are passed through certain polarising' substances, all wave-lengths are filtered out except those which vibrate in one plane; the light is now *polarised*.) Solutions of 'left-handed' molecules rotate a beam of polarised light to the left; the solutions are *laevorotatory*. 'Right-handed' molecules deflect the beam to the right (*dextrorotatory*). This ability of solutions to rotate beams of polarised light is called *optical activity* and it does not occur in solutions from non-living sources because these always contain equal mixtures of right- and left-hand molecules. Living organisms are peculiar in always using one form only of a particular molecule: proteins contain laevorotatory or L-amino-acids, while sugars are always dextrorotatory or D-sugars. The presence of optically active substances is a good indication that they come from a living source.

The reason why living organisms originally adopted left-or right-handed molecules is difficult to guess. Once the choice was made, however, there are sound reasons for sticking to it. When amino-acids chain together to form polypeptides, they spontaneously coil into a helix. If the amino-acids are all of the same isomer, the helix grows rapidly; if they are not, the molecules do not fit together without strain and the helix fails. The same principles apply to the formation of a DNA polymer: a mixture of sugar isomers does not result in a stable double-helix.

Since living things are interdependent and exchange molecules, it was essential that they all adopted the same isomers. Selection would have suppressed whichever non-conforming molecules were in the minority.

THE FIRST CELLS

The first organisms which were fully 'alive' were probably bacteria-like creatures. They would have absorbed energy-rich chemicals from the soup and *fermented* them (that is, respired them without oxygen) to release energy, using the materials to grow and divide. Such creatures which use food manufactured elsewhere are called *heterotrophs*. As bacteria became more efficient in their exploitation of the food around them, they probably began to consume it faster than it was being synthesised by chemical means.

Photosynthesis

A general food-shortage would be a powerful selective-pressure in favour of those organisms which could develop alternative methods of feeding. An important source of energy which was not too destructive for living things to utilise, was sunlight. Certain microbes evolved pigments which were able to absorb sunlight, and used the light energy for synthesis of food molecules. So photosynthesis began; this is the process whereby sunlight induces a reaction involving carbon dioxide and water. Sugar (a store of chemical energy) and oxygen are ultimately produced. (See *The cell concept,* chapter 4 and *Metabolism, movement and control,* chapter 5.) Sugars are a fundamental food from which many others can be synthesised, so such photosynthesisers were among the first *autotrophs*, or nutritionally self-sufficient organisms. (Other autotrophs make carbohydrates by *chemosynthesis*: see *The cell concept,* p. 92.)

Micro-fossils

Relatively large and obvious fossils are found in rocks as old as those of the Cambrian period, between 500 and 600 million years ago but, until recently, very few were known from rocks older than that. For a long time it seemed as though the major phyla of plants and animals appeared, fully fledged, in the early Cambrian. Even supposing that more ancient organisms had existed, the chance of finding fossils intact seemed remote because few sedimentary rocks remained from earlier times. In the 1960s, however, micro-palaentologists succeeded in obtaining remarkable evidence of life in South African rocks 3200 million years old. The fossilised microbes resemble bacteria and *blue-green algae,* the most primitive kind of plant alive today. These fossils represent the oldest living organisms so far identified, yet even these had already diversified into heterotrophs and autotrophs of different kinds.

For the next 2400 million years, the only known fossils are of microbes. They obviously developed a vast range of biochemical abilities, but all contained within a single, undivided body.

Aerobic respiration

As the photosynthetic algae proliferated, they emitted as a waste product quantities of free oxygen, lethal to primitive living things. In the face of selective pressure, those organisms would have survived which could cope with higher oxygen levels. The metabolic trick that evolved was to use the potentially dangerous oxidising ability of free oxygen to oxidise food in a controlled fashion. Rapid oxidation is equivalent to burning and would be destructive in a living cell. Molecules such as the *cytochromes* evolved; these carry out the series of controlled oxidation-reductions which are essential to the efficient metabolism of food with oxygen. This change from *anaerobic* (without oxygen) to *aerobic* (with oxygen) respiration must have been the last really fundamental step in biochemical evolution, and a very successful one, since today only a few bacteria remain obliged to avoid oxygen and respire anaerobically. Efficient photosynthesis demands highly organised membranes to enclose the pigments and to order the enzymes involved; such structures are called chloroplasts. Likewise, efficient aerobic respiration requires membraneous structures which maintain a high concentration of the complex reactants in the correct order; these are mitochondria. The evolution of photosynthesis and aerobic respiration therefore coincided with the evolution of cells with much more complex internal organisation: Today, mitochondria are universally present in the cells of living animals and plants, while chloroplasts occur only in plant cells.

Symbiotic relationships

There is a fascinating theory to explain the appearance of separate organelles such as mitochondria and chloroplasts within cells. Individually, each mitochondrion resembles a bacterium, and in the cells of plants and animals they reproduce independently of the division of the cell to which they belong. Moreover, mitochondria possess genes of their own which they exchange with each other in an unorthodox genetic system. Chloroplasts behave similarly, and resemble blue-green algae whose ancestors were probably the first photosynthesisers on earth. The theory which is now gaining acceptance is that primitive microbes without the ability to respire aerobically absorbed .whole some that could. The lodgers were maintained alive, presumably deriving protection by living within the larger cell, in a mutually beneficial *symbiosis*. The bacteria-turned-mitochondria multiplied independently, and, supplied with food-material and oxygen via the host cells, respired to release sufficient energy for themselves and their hosts. Cells which took in only bacteria gave rise to animals, while those which absorbed both bacteria and blue-green algae developed into green plants.

Cilia and flagella

Lynn Margulis, one of the foremost advocates of this symbiosis theory, believes that other primitive organisms may also have become incorporated into larger cells. *Flagella* and *cilia* (long and short hair-like projections respectively) are commonly found both on the outside of single-celled organisms and on the cell-surfaces of multicellular organisms. They either propel the cell by their motion, or else propel things past the cell. Their basic structure is remarkably uniform whether they come from ferns or human beings. Even more striking is the fact that the structure of the base of cilia and flagella (an arrangement of nine paired tubes in a circle) closely resembles that of the centrioles which play a part in the mitosis of the chromosomes of 'higher' cells (see p. 34). Lynn Margulis has accumulated evidence that cilia, flagella and centrioles are derived from common ancestors which were free-living motile bacteria. Modern representatives of this group are called *spirochaetes*. Certain primitive cells perhaps acquired spirochaetes as surface-symbionts and thereby became able to move more rapidly to areas of high food-concentration. Some spirochaetes which penetrated into the cells apparently came to be converted into the centrioles and mitotic spindle which are responsible for the equal division of chromosomes (Figure 91).

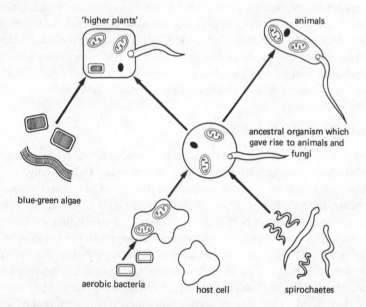

Figure 91   The symbiotic origin of animals and plants
The symbiotic theory proposes that symbiotic union between different kinds of relatively simple cells gave rise to complex cells of the green plants, fungi and animals

Cell-division

Mitotic chromosomal division, with its equal separation of genes, was a crucial stage in evolution. Without it, cells would not have been able to proliferate in a way which could give rise to multicellular organisms. Also, the mechanism of meiosis could hardly have evolved before mitosis, and meiosis is essential for the segregation of genes into eggs and sperm and all the diversity permitted by the reshuffling of genes in sexual reproduction. Meiosis meant a much greater variability on which natural selection could work. Mitosis did not develop to perfection in an evolutionary straight-line, however. There must have been many 'experiments' which failed and today there exists a variety of microscopic organisms which employ unusual forms of mitosis.

MULTICELLULAR ORGANISATION

Single-celled organisms were, and still are, widespread. Many are biochemically extremely versatile, being able to synthesise everything they need from a few simple nutrients. Together, micro-organisms occupy a vast variety of niches, some of which are highly specialised. One might wonder why multicellular organisms evolved.

Cells with organelles tend to be much larger than the bacterial-type cells from which, according to Margulis, they were derived, and size-increase has a number of advantages. A larger cell is less affected by changes in its outside environment than is a small one: it gains greater internal stability. There is a limit to the size-increase possible for a single cell, however. As a spherical body doubles its diameter, its surface area quadruples and its volume increases eightfold. Consequently, in a similarly shaped creature which relies on diffusion through its skin to obtain oxygen and nutrients and to dispose of its excretory products, the diffusion pathway becomes longer. Diffusion is therefore slower, and the increase in available respiratory surface does not keep pace with the increased demand for oxygen. The size-limits imposed by the operation of these physical principles were circumvented by certain cells which aggregated together to form the first multicellular organisms. The cells remained individually small, yet the organism was larger. Even more importantly, multicells exhibit a new level of organisation. The cells of which they consist are not identical, but have different, specialised functions. A cell can evolve a higher degree of perfection as an external protector if it does not simultaneously have to be an efficient excretory organ. This cellular *division of labour* was the chief factor in the evolution of the multicellular organisms.

The evolution of multicellular organisms probably occurred between 800 and 600 million years ago. It may have been made possible when the atmospheric oxygen-level, gradually increasing as photosynthesis proceeded, reached about one per cent of its present level. This would have allowed more efficient respiration than before, and hence the extraction of more energy from a certain amount of food, facilitating the maintenance of larger bodies.

Metazoan animals

All multicellular animals except sponges (whose cells retain a measure of independence) are *metazoans*. There has been a gradual trend in metazoan evolution towards greater division of labour between groups of cells. Similar sorts of cells gathered together are called tissues and different tissues integrated to perform distinct functions are organs. More specialised organs appeared in those phyla which evolved most recently.

Jellyfish, sea-anemones and corals, called *coelenterates*, represent a simple metazoan design. These animals are hollow-bodied with a ring of tentacles, which capture food, around the mouth. Coelenterates are unusual in the animal kingdom in that they are *radially-symmetrical* (circular). This shape admirably suits an animal which is either fixed, or else drifts randomly in the water. Other animals are all *bilaterally symmetrical*, with distinct head and rear ends and two mirror-image halves if divided vertically and longitudinally (i.e. sagittally).

Flatworms are simple bilaterally symmetrical animals which propel themselves by the action of many cilia. Like the coelenterates, they are restricted by the fact that every cell must be close to the surface in order to respire. Coelenterates are hollow and thin-walled which allows them to meet this demand whilst growing quite large. Solid-bodied flatworms are restricted in size and must remain very thin to facilitate diffusion.

The coelom

Other animal phyla were able to diversify as a result of the evolution of the *coelom*. This is a fluid-filled body-cavity which originally conferred rigidity (as in a blown-up balloon), allowed the gut to move independently of the body-wall and provided a space in which organs developed. In *annelid worms*, the group to which modern earthworms belong, the coelom is sub-divided internally into self-contained segments. Each segment contains its own set of organs, co-ordinated by a nerve-cord running the whole length of the body. By the time the annelids evolved, the effects of the size-limit imposed by the rate of diffusion of oxygen for respiration were being altered. Respiratory exchange (that is, the exchange of oxygen from the water outside for carbon dioxide created by respiration within the animal) is facilitated if the exchange-surface is enlarged and if the water or the body fluids move. Some annelids developed large, thin gills which absorbed oxygen and released carbon dioxide readily, and a blood-system which distributed the gases to and from the tissues more rapidly than diffusion could. Such animals can grow larger than flatworms.

The annelid design was modified in the *arthropods*, animals whose segments developed legs and which acquired hardened skins or *exoskeletons*. Legs allow faster movement than occurs in legless annelids, and a tough skin is an important pre-requirement for life on land, a habitat which was later adopted by many arthropods. The arthropods form one of the largest phyla; they include crustaceans (such as the crab), insects and spiders.

The annelid-design also underwent modification in a different group of organisms, which led to the phylum *echinoderms* (which includes starfish and sea-

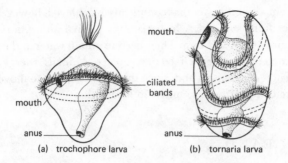

mouth

ciliated
bands

mouth

anus

anus

(a) trochophore larva   (b) tornaria larva

Figure 92   Trochophore and tornaria larvae
Annelids and molluscs have similar trochophore larvae while echinoderms and less advanced chordates have tornaria larvae. Larval similarities denote evolutionary kinship

urchins) and the *chordates* (whose most obvious living representatives are the backboned vertebrates). These two phyla seem widely different, but their common origin was established by detailed study of their *larvae* (free-swimming young stages) (Figure 92). Vertebrates have no such larvae, but primitive chordates do, and all chordates are alike in the possession of a rigid rod or *notochord* along the back. The evolution of the vertebrate sub-phylum from the more primitive chordates was the only really radical change in animal design that occurred after pre-Cambrian times.

Fungi

Many authorities classify the *fungi* as a separate kingdom alongside those of the plants and the animals. Fungi adopted the rôle of feeding on dead organic matter as a source of energy, and never possessed (or else lost) the ability to photosynthesise. There is today a great variety of fungi; their characteristic organisation is a collection of filaments forming a meshwork, parts of which may make up very complex structures.

Plants

Algae
The first photosynthetic plants with organelles were the *algae*, and many one-celled algae persist today. Other algae grew into multicellular filaments and in some the filaments branched and formed large leaf-like structures, as in sea-weeds. Present-day algae exhibit a vast variety of forms and reproductive cycles, and between them they carry out most of the earth's photosynthesis.

'Higher' plants
Those multicellular plants which evolved from green algae and spread to the land are the bryophytes and the vascular plants. As in the animals, the trend towards size-increase has been accompanied by an increase in complexity. Plants

differ from animals in the way that complexity is achieved, however. Since their energy comes from sunlight, which requires passive absorption, their aerial surfaces are mainly flat and thin. Their need to obtain water and nutrients from the soil is related to their being fixed in one place. Animals' energy derives from the bodies of other organisms and, since this requires them to move about, they develop more compact bodies, suited to locomotion.

LIFE ON LAND

Life originated in water and, throughout most of its evolution, remained there. Water is buoyant, is essential to the maintenance of living cells and probably provided some protection from the damaging effects of ultra-violet light. Evidence suggests that when the accumulation of oxygen from photosynthesis reached about ten per cent of the present level, the high altitude screen of tri-oxygen so created would have begun to shield the earth from ultra-violet radiation. It might have been at this time that some living organisms, always in competition for the available niches, emerged to colonise the land. The plants had achieved this major evolutionary advance by the Silurian period, about 425 million years ago, and probably earlier; they were the first organisms to do so.

Air and water have very different properties, so that land organisms, surrounded by air, have evolved different adaptations from those of their aquatic ancestors. It is essential that living organisms retain within them a certain percentage of water, since all the chemical reactions necessary to life occur in solution. Organisms with moist surfaces can tolerate exposure to air only under very damp conditions, otherwise water evaporates from their cells faster than it can be replaced. Organisms which evolved relatively watertight cuticles or skins were able to colonise dry habitats.

Terrestrial respiration

The respiratory surface, where carbon-dioxide is exchanged for oxygen in aerobically-respiring organisms, must be thin-skinned for rapid gas-diffusion. Such a surface cannot be both sufficiently thin and watertight. Land animals evolved respiratory surfaces as invaginations, or pockets, rather than outgrowths (such as gills) which are characteristic of aquatic animals. Such surfaces suffer relatively little water-loss by evaporation since their openings to the exterior are restricted and the air within them is moist. Thus vertebrates evolved lungs and insects evolved *tracheae* (branching tubes leading from the outside into the body).

Air contains 210 cm³ of oxygen per litre, while a litre of sea-water contains only about 5 cm³ and one of freshwater about 7 cm³. Less air than water therefore needs to be processed in order to extract the same quantity of oxygen. In addition, water is denser than air, so that pumping it over gills requires more energy than pumping air in and out of lungs. These factors contribute to the fact that certain terrestrial air-breathing animals have evolved the ability to fly; this requires large amounts of energy and hence rapid and efficient gas-exchange.

Support

Aquatic organisms are supported by the buoyancy of water, while land creatures would tend to collapse without a supporting skeleton. In plants, support is provided by the turgor of cells and by the rigidity of woody tissues. Different groups of animals have evolved external skeletons (as in arthropods) and internal ones (as in vertebrates). In these cases, the skeleton probably evolved in response to selective pressures operating in the aquatic environment before land was colonised. Thus, arthropod armour conferred protection against predation and the vertebrate internal skeleton acted as a system of levers to which locomotory muscles were attached. On land, existing skeletons acquired new, supporting functions.

Other adaptations

Only a few of the adaptations necessary for terrestrial life have been discussed here. There are many others, particularly physiological ones, which you should consider, bearing in mind the very different properties of air and water. Those organisms which became totally terrestrial were those which evolved the ability to reproduce without dependence on water. Reflection on the mechanisms of reproduction discussed in chapter 8 should give you ideas on how this came about.

Adaptive radiation

Once living organisms had begun to exploit the marginal water-land habitats, powerful new selective pressures would have exerted their influence. When adaptation to the new environment had reached a certain level, that is, when the main body functions were able to occur under atmospheric conditions, the relatively rapid specialisation of different types into different niches, or adaptive radiation began. Fossils demonstrate that land plants, for example, underwent a radiation in the late Silurian and early Devonian periods and developed highly diversified structures. Discovering true evolutionary connections between fossil organisms which date from such radiations is a difficult task. Frequently, two unrelated plants or animals will develop similar adaptations in response to similar selective pressures. These similarities are said to be the result of convergent evolution and not common ancestry.

THE ORIGIN OF THE LARGER TAXA

All the major phyla were established by the time multi-cellular forms appear in the fossil record at the beginning of the Cambrian period. This has led to speculation that the larger taxa, such as phyla and classes, evolved by a process other than that of the accumulation of small changes in populations that accounts for the origin of species by natural selection. Today, although the details are still not fully understood, most scientists believe that the processes involved in species-formation and phyla-formation are essentially similar.

At the inception of a new phylum, the number of individuals involved may be small and they may evolve rapidly and contribute relatively little to the fossil record. Rates of evolution do vary considerably according to the changes in selective-pressures. The easiest way to estimate rate of evolution among fossils is to measure the size-change of an organ over a period of time. The unit used to measure this is the *darwin*, where one darwin is a change in dimension of one per cent per 10000 years.

No organisms intermediate between two phyla are known from the fossil-record, although we presume that common ancestors existed. In more recent times, though, there is good fossil evidence for intermediates between classes and between orders. The *creodonts*, for instance, were an early form of carnivore which linked modern carnivores with their insectivore ancestors.

The line which evolves into the more 'advanced' group commonly arises early in its ancestors' history. In the well-known amphibian-reptile-mammal sequence, for instance, the line that led to the reptiles appeared almost as soon as the amphibians evolved, and mammal-like reptiles arose as soon as the reptiles were established. This partly accounts for the large discontinuities between modern forms with common ancestors in the past; the modern forms are the result of much longer separate evolutionary histories than appearances suggest.

This pattern of evolution is important as it means that the group which eventually evolves to a new level of organisation does not develop the later-acquired, specialised characters of the group from which it is derived. It is the relatively generalised structure and function which apparently makes further evolutionary change possible. If an organism becomes very specialised in a particular direction it may not be able to change radically in response to new circumstances.

QUESTIONS

1 Summarise events which may have led to the evolution of life. What are the weakest hypotheses in your account? Suggest possible alternative explanations.

2 Choose a group of plants or animals which you know to have become extinct and suggest reasons for their extinction.

3 What does 'adaptive radiation' mean? Give examples of adaptive radiation within a phylum, a class and a genus.

4 What structural and physiological changes must aquatic plants have undergone in order to evolve into land-living forms? How do these changes differ from those undergone by animals which adapted to terrestrial life?

12 Human Evolution

Within the last twenty years, fresh fossil and archaeological evidence has given us new insights into the fascinating problem of the evolution of man. It is a unique story: our species appeared about 35000 years ago and, in this relatively short time, has multiplied at an accelerating rate and come to dominate the earth's resources. Such rapid change could not have been brought about by biological evolution alone, but resulted from the evolution of 'culture' to an extent that is unique to humans. Culture is the social, learned component of existence which an individual acquires from his group. It allows great flexibility of behaviour, rapid adaptation to changed conditions and innovation of ideas which can be shared. The importance of culture relies on man's capacity for learning, which is far in excess of that of other animals.

The human line has been traced back seventy-five million years to small mammals: rat-sized, insect-eating *tree-shrews*, some of which survive today, relatively unchanged. Others, in response to subtle and unknown environmental influences, developed longer fingers with flat nails instead of claws and improved their ability to leap and cling. The three-dimensional arboreal life depended largely on sight, and smell became relatively unimportant. Tree-shrews developed larger eyes which came to point forwards so that the fields of view overlapped and gave binocular vision, a great improvement for judging distances. Simultaneously, the snout receded. By about sixty-eight million years B.P. (before present), the tree-shrews had altered sufficiently to be called *primates*, and early representatives are thought to resemble modern bush-babies, tarsiers, lemurs and lorises. Tree-life is varied and physically precarious, so that quick reactions are at a premium. As primates evolved, their brains became better at synthesising and storing information. Forty million years B.P., the monkeys arose and, from them, the apes, man's closest relatives (Figure 93).

The publication of Darwin's *The Descent of Man* in 1871 first put forward the case for the descent of man from apes. Then there was virtually no fossil evidence to support the proposal, but the search for the 'missing link' between apes and man soon began. Darwin knew of one fossil skull, found in the Neander *thal* (valley) in Germany, which appeared to belong to a primitive human; other Neanderthal fossils followed. In 1891, yet more primitive skull-fragments and teeth were discovered in Java by a Dutchman, Eugene Dubois. These were thought to be ¾ million years old and to belong to a different species of man now known as *Homo erectus*. Exciting though these finds were, they did not solve the problem of where the human genus, *Homo*, came from. Considerable numbers of man-like fossils have also been found in Eastern China and may be the same species as *H. erectus*.

Tree shrew

Lemurs

Monkeys

Ape

Figure 93

AUSTRALOPITHECINES

Progress was made in 1924 when a skull was discovered at Taung in Botswana, and sent to Raymond Dart, Professor of Anatomy at Johannesburg University. It had a brain-size of about 500 cm³, within the range of apes' brains (modern humans average 1400 cm³) but its jaws and teeth were human-like. Moreover, the *foramen magnum*, the hole at the base of the skull which admits the spinal cord, pointed downwards rather than back, indicating that the creature stood erect. Dart announced to a disbelieving world that he had found a hominid (a member of the human-line). He named it *Australopithecus africanus*. Ten years later, Dart's supporter, Robert Broom, a distinguished South African palaeontologist, discovered more australopithecine fossils in South Africa and suggested that they were two million years old.

In 1959, Louis and Mary Leakey were working in Olduvai Gorge, a dry river valley in Tanzania, East Africa. They discovered a larger kind of australopithecine skull in the sedimentary strata, dated at 1.75 million years old. Later, smaller hominids were found, including some that were sufficiently man-like for Leakey to name *Homo habilis*. In an attempt to learn more about the hominids, an international expedition to the Omo valley in Ethiopia was organised in 1967. Omo resembles Olduvai in being part of the Rift Valley, although its river still flows into Lake Rudolph. It has the advantage that exposed strata are 2000 feet thick and span a longer and earlier time interval than do the Olduvai strata. Moreover, layers of volcanic ash can be dated by potassium-argon methods, so dating the fossils sandwiched between the layers. Valuable hominid fossils, discovered from 3.7 to 1.8 million years old, were discovered. Meanwhile, the Leakeys' son, Richard, discovered even more fruitful fossil beds on the eastern side of Lake Rudolph, and a picture of the population of large australopithecines began to emerge. It seemed that Dart had been right; the australopithecines were truly hominid and, while the large vegetarian form had become extinct, the smaller *Australopithecus africanus* (or a relative – opinions differ) evolved into *Homo*, true man.

Australopithecines were thought to occupy such an important position in man's evolutionary history that much effort was expended on reconstructions of their way of life.

It must be stressed that this interpretation of the lifestyle of australopithecines and of their mode of evolution is speculative – informed guesswork on the part of experts. There are other equally defensible viewpoints. The angle of the australopithecine head and the shape of the pelvic girdle are evidence of an upright posture and two-legged (*bipedal*) walking. The fact that their presumed ancestors, apes, swung by their arms was important in the evolution of bipedalism. Apes were already partially erect and could reach out and manipulate objects with their hands whilst sitting. For reasons we shall never fully understand, one type of ape apparently took to spending more time on the open savannah at the edges of the forest. With hands already adapted to grasping, selection would perhaps have favoured a more upright posture which facilitated carrying food. The diet of a savannah-living australopithecine would have been different from that of a tree-living, fruit-eating ape. Cereal grains must have been available from the abundant grasses, and the mastication of these requires large grinding molars moved in a rotary fashion. Such side-to-side grinding would be facilitated if the large canines, characteristic of apes, were reduced. Canines are weapons of attack and defence, as well as teeth for feeding, and their reduction may have made australopithecines vulnerable. Here, their increasing manipulative ability would have been a tremendous advantage; the ability to throw a stone accurately might have compensated for small teeth.

Tools

Tool-use (and any object, even a stone, which extends the hand's abilities is classed as a tool in this context) is fundamental to all hominid behaviour. There is argument as to whether tool-use pre-dated bipedal walking or vice-versa, but the two must have evolved together, alongside increases in the size and complexity of the brain.

Mary Leakey has investigated many stone tools from Olduvai, dated between one and two million years old. She has identified choppers, hand-axes, cutting-tools and anvils – evidence that the australopithecines not only used tools, but that they made them. Killing large game is risky and difficult for a small animal (*Australopithecus africanus* stood about 1.37 m (4.5 ft) tall) and could have been achieved only through a co-operative effort. Hunting in bands and food-sharing after a kill probably stimulated the growth of social interaction and the accompanying need to communicate. Selection would have favoured brains capable of handling increasingly elaborate data.

Opinions differ as to whether australopithecines used true language: it is probable that they relied on a sophisticated system of gestural or 'body' language, facial expressions and calls.

Behaviour of living apes

Research into ape-behaviour in the wild has thrown light on to the probable behaviour of human ancestors. Jane Goodall, at one time secretary to Louis Leakey, studied a wild troop of chimpanzees at the Gombe Stream in Tanzania. For many years she camped near the troop, became accepted by them, and made observations which revolutionised our picture of chimpanzee behaviour. The use of tools, it seems, is not the prerogative of man. The Gombe chimps strip leaves from twigs, for example, insert them into termite heaps and eat the termites which cling to the twig when it is withdrawn. They hunt small animals in groups and show complex patterns of non-verbal communication, some of which resemble human gestures. Chimpanzees are extremely inventive, and incidents at Gombe stream illustrate how increased intelligence might have selective advantage. Male chimps make noisy charges when they are excited, but the charge of a subordinate is usually ignored by a dominant animal. One chimp, named Mike by Jane Goodall, had a low position in the hierarchy of adult males. Empty paraffin cans around the camp attracted him, and one day he collected some, and hurled them before him in a dramatic charge on the troop's dominant males. This attack was deliberate and very effective: the chimpanzees scattered, terrified by the rolling cans. After Mike had repeated his charges several times he was sufficiently respected to gain a new position as dominant chimpanzee, one he retained for many years after the confiscation of the cans. The situation is unnatural in that tin-cans are not normally available to wild chimpanzees, but it illustrates that a particularly intelligent animal may gain an advantage over his peers which may contribute to his survival. Dominant chimps enjoy many advantages, one of which is the right to mate preferentially and so to pass their 'advantageous' genes on to the next generation.

Modern assessment of Australopithecus

The notion that *Australopithecus* was man's direct ancestor has been challenged recently. In 1972, Richard Leakey discovered a skull in northern Kenya which was dated at 2.1 million years old and yet had a brain-capacity of 800 cm³. Australopithecine brains seldom exceeded 550 cm³. Leakey's find, known as skull 1470, was assigned to the genus *Homo*. Since then, rich finds of large-brained hominids have been made in Ethiopia and dated at around three million years B.P. It seems that *Homo* has a much longer history than was previously supposed, and perhaps evolved alongside the australopithecines for four million years. The search for the point at which they diverged has been pushed further back in time. Australopithecines nevertheless remain important as an illustration of a stage through which the hominids evolved, even if the direct line of man's ancestry lies elsewhere.

MODERN MEN

Hominid fossils are more abundant and more easily-interpreted in younger rocks, so that much more is understood about our more recent ancestors. Probably, modern man is a descendant of *Homo erectus*, a species which first appeared about 1½ million years ago. Around 300000 years ago, anatomists consider that an invisible, arbitrary barrier was passed and that *H. erectus* evolved into *Homo sapiens*.

There were several sub-species of *H. sapiens*, a wide spread one of which was *H. sapiens neanderthalis* (Neanderthal man) a large-brained game-hunter.

The oldest remains of *H. sapiens sapiens*, the sub-species to which we belong, appeared around 35 000 years ago, probably having evolved from Neanderthal man. Men then were as large-brained as they are today, and existed as hunter-gatherers in co-operative bands.

The Palaeolithic era

When hominids began to fashion stone tools, about two million years ago, an era known archaeologically as the *Palaeolithic*, or Old Stone Age, began. Throughout the Palaeolithic, our ancestors were hunters and gatherers of wild plants. Probably, hunting gave Palaeolithic people a complex social system, but the interpretation which follows is necessary speculative. Related males began to form hunting bands, leaving their females and children at a 'base' or 'home'. Eventually, each male remained permanently associated with the female who bore his children; such 'pair-bonding' was probably linked with the development of year-round sexual-receptivity in the female. Mental attributes such as curiosity and long-term motivation must have developed, and individuals would have come to rely less on genetically-controlled behaviour and more on learned rules and customs. It may be that the design of the human nervous system makes it easy to induce feelings of guilt when known rules are broken. Perhaps 'conscience' is essential to human society.

Language

A richly complex social system probably evolved rapidly with the development of true vocal language which allowed discussion and description of emotion and ideas. Language is tremendously important in thinking, planning, problem-solving and co-operation. It is a unique system of communication in that it allows totally original statements to be made and understood, and that it is symbolic so that abstract ideas can be communicated. Language has to be learned, and the prolonged human childhood is a time when language and the other intricacies of social behaviour are acquired. Changes in brain-organisation must have occurred as language evolved, and it seems that brain-size, particularly that of the *cortex* (the outer layer of the brain where the 'higher' thinking processes occur) increased at this time. These factors probably produced enhanced intelligence.

Chomsky, a famous linguist, has demonstrated that all modern languages have fundamental similarities of structure, or grammar. This implies that they originated from a common language which evolved only once and diversified later.

It is interesting to consider whether man's closest living relatives, the apes, have any capacity for language. Apes' vocal apparatus is not capable of speech, so two American research workers. Allan and Beatrice Gardner, reared a young female chimpanzee, called Washoe, and taught her the American deaf-and-dumb sign-language which uses hand-gestures. Chimpanzees are manually dexterous and good mimics. Washoe learned the signs for numerous objects and concepts and would string them together into original 'sentences'; she even invented signs of her own. Although apes do not symbolise like this in the wild, they clearly have the capacity to cope with the abstraction necessary for a simple language.

The Neolithic Revolution

The Palaeolithic food-gathering economy was gradually replaced by one of food-production in the *Neolithic*, or New Stone Age, which began about 8000 BC. The term 'Neolithic' implies a cultural level, not a period of time. There persist today a few small groups of hunter-gatherers whose level of organisation has not yet reached the Neolithic. The Neolithic was a true 'revolution' in life-style, although the transformation was gradual – a cultural evolution. It began in the Middle East between Greece and Western Iran (Figure 94). Wild wheat and barley grew in these areas and were presumably collected for their grain. Possibly spilt grain germinated accidentally at the settlement, resulting in a crop close at hand and demonstrating that seed could be sown deliberately where it was needed. The cultivation of wheat and barley became common practice and supplemented hunting, gathering and fishing.

Grain was fundamental to the new way of life; its nutritional rewards were high for the labour involved and it could be stored easily for use all the year. The evolution of modern cereals from Neolithic wild grasses is partially understood. It has involved mutations and chromosome doublings allied to the artificial selection applied by man in choosing the best and biggest ears to provide seed for the next year's crop. Carbonised grains, impressions of ears of cereal and fossil

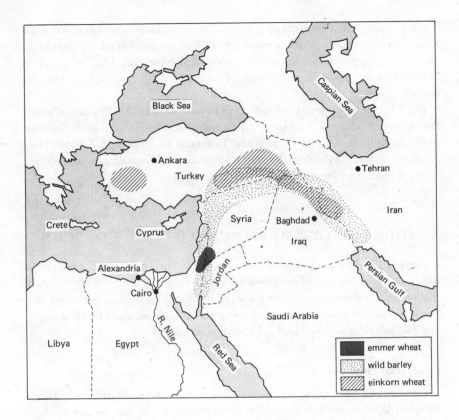

Figure 94   Site of origin of the Neolithic Revolution
The centre for the Neolithic Revolution corresponded to an area where wild wheat
and barley occurred and still occur. Their presence made early agriculture possible.

tools have been recovered from Neolithic settlements and give a good picture of
the agriculture.

Wild ancestors of the goat and sheep lived in the same areas and were
probably the first animals domesticated for food, followed by cattle and pigs.
Mixed farming began early; animals grazed the stubble after reaping and
provided manure which improved soil-fertility. Selective breeding by Neolithic
man would have meant retaining the largest, most docile animals and discarding
the scrawny and intractable. Animals were kept initially for meat and hides; as
they became tamer they were milked and used to carry packs and pull ploughs.

The development of civilisation depended on the gradual transition from
hunting to farming. Food-production could be partially controlled for the first
time and so an increasing human population could be fed. Instead of small,
scattered bands of nomadic hunters there developed larger, settled communities.
Neolithic people made permanent houses, often sited together within fortified
villages. The oldest-known settlement large enough to be called a town grew out
of villages at Jericho between 8000 and 6800 BC. By 6800 BC, at Çatal Hüyük, in

Turkey, there was a flourishing town whose inhabitants were occupied with agriculture and trade. With a more secure and settled way of life, Neolithic people had opportunities to develop their craftsmanship. The invention of pottery was an important stride forward for food-storage and cooking and also for art.

Neolithic culture spread gradually from its nucleus in the Middle East towards the Atlantic shores of Europe. As it developed, populations with different cultural histories must have met and interbred and exchanged ideas. The mechanism was established for the transmission of information to produce cultural change at an ever-accelerating rate.

A MOLECULAR APPROACH TO MAN'S DESCENT

Experts are in constant disagreement as to which ancestral branch gave rise to which. Some prefer to give most weight to fossil evidence, while others maintain that behavioural comparisons are more important. Objective comparisons of the evolutionary 'distance' between living species can now be made biochemically, and this provides a valuable third line of evidence.

DNA differences

The nature of a species depends upon the coded information carried by its DNA. In time, DNA-mutations gradually accumulate so that the species changes phenotypically and is said to evolve. The longer this process continues, the greater will be the number of differences in the DNA. If the DNA of two living species is compared, the evolutionary distance between them can be expressed in terms of the number of DNA differences.

Elaborate techniques enable biochemists to separate the two strands of a DNA double-helix. A single DNA strand of, say, a man, is then induced to recombine with a single strand from an ape. Linkage of the strands will occur except at the points where the links are chemically different owing to mutations. The percentage of linkage-failures between man and chimpanzee is about 2.5 per cent; between man and monkey more than 10 per cent. This confirms that the human-line separated more recently from the apes' than from the monkeys'.

The immunology method

Another approach to biochemical differences between living species relies on an animal's capacity to manufacture *antibodies*, chemicals which react with foreign molecules they encounter. If a particular human blood-protein is injected into a rabbit, the rabbit produces antibodies which react specifically with that protein and no other: this is called the *immune reaction*. *Serum* (blood with cells and protein removed) from a rabbit which has been so injected will produce a large precipitate if it is mixed with the original human protein. Identical rabbit serum mixed with a comparable protein taken from a different animal produces a

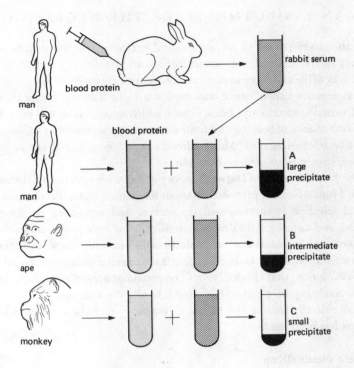

Figure 95 The immunology method of establishing evolutionary relationships
A blood-protein from a human is injected into a rabbit. This rabbit's serum mixed
with more human blood-protein produces a precipitate, (A). Serum from the same
rabbit mixed with blood-protein from an ape produces more precipitate, (B), than
when mixed with monkey blood-protein, (C). This demonstrates a closer
evolutionary relationship between men and apes than between men and monkeys

different degree of precipitation depending on the similarity between the two
blood-proteins. Chimpanzee-protein produces a large precipitate in human-
sensitive rabbit-serum, while less precipitate forms from monkey-protein (Figure
95). The greater difference between human and monkey proteins again indicates
the earlier separation of their *lineages* (lines of descent) than those of apes and
men.

Such molecular methods are important, because they provide a standard
method of comparison of evolutionary distances. It is unfortunate that fossils
cannot be subjected to the tests! The time at which certain lineages separated has
been calculated from the biochemical differences between their living representatives. Such calculations assume that the rate of evolutionary change (or
substitution of one DNA unit for another) remains constant – an assumption
which may not always be justified. However, as methods of matching molecular
data to palaeontological data improve, we shall gain greater insights into our
own line of descent.

HUMAN EVOLUTION SINCE THE NEOLITHIC

Until the development of an agricultural economy, the maximum human lifespan, rarely attained, was about forty years. Neolithic people, with their settled way of life and more abundant food, frequently survived to a greater age. Natural selection can influence only those traits which appear before the end of an individual's reproductive life: a disease which appears later will not affect its possessor's chance of bearing children, so that a genetic tendency to the disease cannot be selected against. Many ailments thus appear in old age in modern societies; cancer is an obvious example.

After the publication of Darwin's theory of evolution by natural selection, his cousin, Francis Galton, proposed methods for human biological improvement. Natural selection is short-sighted: to survive and reproduce is biologically essential, and failure means death or sterility. Men have produced spectacular improvements in plants and animals by allowing only those with the best characteristics to breed. Galton suggested that similar principles should guide human evolution; that children should be produced according to their parents' merits. Such *eugenic* programmes could hardly be compulsory but, on the negative side of avoiding the births of handicapped babies, much voluntary progress has been made.

Genetic counselling

Studying the family pedigrees allows a genetic-counsellor to make predictions about the risk to prospective parents of an abnormal child. There are two main areas in which successful predictions can be made. The first concerns analysis of human chromosomes.

Chromosome analysis

If a family-history shows a tendency to chromosome-abnormality, examination of the chromosomes of their white blood-cells may reveal whether a couple are likely to have affected children.

In sex-linked diseases, such as haemophilia, the defective gene is carried on an X-chromosome, (see p. 46). Since the gene's effects are recessive, a heterozygous woman does not show the symptoms of the disease but, on average, half of her sons will be haemophiliac. In this case, it is the chromosomes of the unborn baby which must be analysed. To avoid damaging the foetus itself, a technique called *amniocentesis* was developed. A hypodermic needle is inserted into the mother's uterus through the wall of her abdomen, and withdraws some fluid containing cells from the foetus (Figure 96). These embryonic cells are examined microscopically. The sex of the cells can be determined, since each female cell contains a darkly staining *Barr-body*, absent from males. The Barr-body is one of the two X-chromosomes, tightly condensed. If the baby is a boy, the mother is given the option of having an abortion. The cells may be *cultured* (allowed to multiply under laboratory conditions) for further examinations.

Techniques of amniocentesis and cell-culture are valuable in the search for other chromosomal disorders such as Down's syndrome (see p. 62), and in this

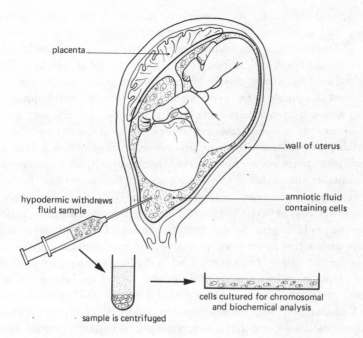

placenta

wall of uterus

hypodermic withdraws
fluid sample

amniotic fluid
containing cells

cells cultured for chromosomal
and biochemical analysis

sample is centrifuged

Figure 96   Amniocentesis
A sample of amniotic fluid containing cells derived from the foetus is withdrawn by
hypodermic syringe. The placenta is located beforehand by ultra-sound and the
foetus is manipulated away from the puncture-site. Ideally, amniocentesis is
performed about sixteen weeks after conception

case may be carried out as routine on older mothers who have a greater risk of
bearing an affected child.

Biochemical disorders

The second kind of prediction involves biochemical analysis. Certain mutant
genes can be detected in heterozygous carriers who do not show symptoms of the
disease, but who risk having a diseased homozygous child. Some haemoglobin
abnormalities can be discovered in this way. The amniotic fluid may be analysed
biochemically; recently a correlation has been demonstrated between the
presence of a high level of a certain protein in the amniotic fluid and the birth to
that mother of babies with the spinal deformity, spina bifida.

Tests may also be performed on the blood or urine of a new-born baby to
identify metabolic disorders. These do not prevent the birth of an affected child,
but they can sometimes avoid the manifestation of the worst phenotypic effects.
The case of PKU, in which the amino-acid phenylalanine must be withheld from
the diet to allow normal development, has already been mentioned (p. 64).

Genetic load

Disease causes physical and emotional distress to sufferers and their families and great medical expense for the community: preventing genetic disorders seems nothing but a blessing. There are long-term evolutionary consequences, however. Our low birth- and death-rates mean that natural selection is removing harmful genes from the population less effectively; we are accumulating a *genetic load* of harmful mutations. Aborting embryos with genetic handicaps seems to circumvent this problem, but even this is not straight-forward. Suppose that a woman who is a haemophilia-carrier wants two children. She may have any male-foetuses she conceives aborted, and persist until she has two daughters. Probability says that one of these, i.e. fifty per cent of the children, is likely to be a heterozygous carrier like her mother. Had the opportunity for amniocentesis not been available, the chance of a heterozygous daughter would have been one in four, twenty-five per cent. Since heterozygous girls are more likely than haemophiliac boys to have children of their own, they are the main distributors of the haemophilia gene to the next generation. The result of widespread selective abortion of males is an increase in the gene-frequency of the mutant gene in the population. The genetic-load in the human-gene-pool is increased.

Since most mutations are harmful initially, one of the best techniques for reducing our genetic-load over a long period of time may be to reduce mutation-rates. Unfortunately, with the development of nuclear power, and its attendant mutagenic radiations, and the ever-increasing list of mutagenic chemicals which are added to our food, water and air, this prospect seems remote.

On the optimistic side, however, it is true that most genes have multiple effects and it may be that genes which appear with unexpectedly high frequency do so because the carriers of the mutant genes have other, undetected advantages. If this view is valid, removing the phenotypic disadvantages of such genes may be in our long-term interest.

Fertility

In pre-Neolithic times, the high human death-rate necessitated a high birth-rate to maintain the population. Natural selection therefore favoured high fertility. In the modern developed world, where most children survive to adulthood, families limit the number of their offspring by choice, so that high fertility is no longer at a premium.

Sperm-banks

Infertile couples who would previously have remained childless may now be enabled to have a child of their own. Commonly, the man's sperm-count is not sufficiently high to make fertilisation likely. This can be remedied simply by allowing him to donate a number of sperm-samples at different times, concentrating them together and artificially inseminating the woman with the concentrate. Sperm may be stored for long periods at sub-zero temperatures without harming their vitality. If the husband is entirely infertile, the couple can choose to have a baby by insemination of a donor's sperm from a *sperm-bank*. Care is taken to ensure that the donor is reasonably similar to the husband, but his identity remains unknown.

Some people believe that such techniques provide a golden opportunity to improve the human race in the way Sir Francis Galton envisaged. Instead of secrecy, they recommend that the details of the donor's physique and character should be known. Prospective parents (perhaps even normally fertile ones) could then choose to have their baby fathered by someone of excellent qualities. To avoid embarrassment and legal difficulties, such a scheme would best operate after the donor' death. This would also give the opportunity to eliminate as donors anyone who suffered from disease or senility in old age. The pros and cons of the idea make an interesting topic for speculation.

QUESTIONS

1 Of what value is the study of living apes in the effort to understand human ancestors?
2 How can you explain the fact that humans, presumably all evolved from one ancestor, occur in distinctive races?
3 Are humans exempt from the forces of natural selection?
4 What is the significance of standing upright in the evolution of man? How do you think the upright stance came about?
5 Write an essay on eugenics.

13 Genetic Engineering

Most revolutionary of all the available methods of directing evolution to human ends is that of *genetic engineering*, or deliberately altering genes. Our knowledge of such engineering comes largely from the study of microbes.

There are two areas of particular interest to genetic engineers. One is agricultural: the production of better-yielding, more disease-resistant crops and stock. The other is medical: the genetic improvement of the human race. Agriculturally, it would be useful, for example, to be able to transfer the genes for nitrogen-fixation from plants such as peas and beans, whose symbiotic root nodule bacteria already possess this facility, to the cereal crops which do not. Cereals that were self-sufficient in their nitrogen-requirements would dispense with the need for expensive and potentially-harmful nitrate-fertiliser. A virus is a suitable carrier for such genes, already perfected by millions of years of evolution as a device for introducing foreign DNA into cells.

Research has not yet reached the stage of transferring selected genes, but viruses have been used to increase a plants' synthesis of a particular amino-acid. Synthetic polyadenylic acid (AAAAA . . . AAA) can be added to the genetic material of the tobacco-mosaic virus (TMV). AAA is the genetic-code for the amino-acid lysine. When the treated TMV is applied to tobacco-plants, the tobacco cells produce larger-than-normal quantities of poly-lysine. Since the modified virus replicates itself, it can be transmitted through several plant-generations. Such selected use of treated viruses could perhaps improve the nutritional value of a crop.

Medically, genetic deficiencies such as phenylketonuria (see p. 64) could be remedied if a method were found for introducing the missing enzyme into the embryo. Some progress has been made in this direction. When the Shope virus, for instance, infects human beings, it produces apparently no symptoms other than a decreased blood-level of the amino-acid arginine. This occurs because the virus induces the synthesis of *arginase*, the enzyme which breaks down arginine. It happens that some people suffer from a disease caused by the failure of the ability to decompose arginine. Excessive quantities of the amino-acid accumulate and result in severe mental retardation, epilepsy and metabolic abnormalities. If the Shope virus proves to be safe in other respects, there are obvious uses for it.

Recombinant DNA

Certain viruses are unable to multiply within some bacteria which they invade because the viral DNA is cut into pieces by bacterial enzymes called *restriction endonucleases*. Now more than twenty restriction endonucleases are known, each of which cleaves DNA at a particular sequence of bases. These enzymes are vital tools of the genetic engineer and are used to cut up DNA in predictable ways. The cut ends are 'sticky' and can be joined again, to each other or to foreign DNA, by other specific enzymes to form *recombinant DNA*. DNA from more than one species can be recombined in this way.

Most bacteria contain freely-replicating, circular lengths of DNA, independent of the main chromosome, called *plasmids*. If restriction endonucleases are used to cut a circular plasmid, fragments of foreign DNA can be inserted and the 'ends' of the plasmid rejoined. The hybrid plasmid may then be re-inserted into its bacterium where it will be replicated normally at each bacterial division. In this way, genes from the toad, *Xenopus laevis*, have replicated within bacterial plasmids. When it becomes possible to induce introduced genes to form proteins, a technique for creating bacteria with human-directed functions will be available.

In early 1977, copies of rat insulin genes were successfully produced within the bacterium *Escherichia coli*. Rat insulin m-RNA is used as a template from which the corresponding DNA is constructed. This process, the reverse of normal transcription, is catalysed by an enzyme called *reverse transcriptase*. The resulting *complementary-DNA (c-DNA)* is a copy of just that part of the insulin gene which is normally transcribed. The m-RNA is removed enzymatically and further treatment with reverse transcriptase forms a double-stranded DNA molecule from the single-stranded c-DNA. Research is now proceeding to stimulate the insulin 'gene' to be transcribed and translated into insulin in its bacterial host.

Work with such bacterial 'recombinant' DNA which contains genes from other organisms is considered particularly hazardous. It is possible to insert 'foreign' DNA into a bacterial chromosome, but impossible to predict exactly how it will behave once the bacterium is inside a complex, living body. There is the fear that bacteria containing foreign genes might escape from the limits of laboratory conditions and come to infect humans, perhaps inducing cancers or other metabolic disturbances. The risk is minimised by using specially weakened strains of bacteria which cannnot live outside laboratory conditions; nevertheless, the common usage of *Escherichia coli*, a normal inhabitant of the human gut, as an experimental organism must give some cause for concern.

In January 1975, the Ashby Committee produced a *Report on the experimental manipulation of the genetic composition of micro-organisms* which recommended precautions to be taken during the conduct of experiments. There is a suggestion from some quarters that certain research should stop altogether until its implications are more thoroughly understood.

Hybridisation techniques

Less alarming than DNA-recombinations are experiments designed to hybridise cells from two different species of plant. In the early 1960s, enzymes were discovered which can dissolve plant cell-walls and liberate the living protoplast within. Protoplasts, from favourable material such as tobacco-leaves, spread (like a suspension of bacteria) on to nutrient agar, will grow and divide. Two kinds of protoplast may be brought together under these conditions and, with the aid of a chemical inducer, fused to form hybrid cells whose nuclei contain four sets of chromosomes. Some hybrid cells proliferate successfully to produce entire plants. This procedure has been tried with petunias with white, long flowers and red, short flowers. The hybrids' flowers were medium-length and pink, like those of the hybrids of normal cross-pollination. So far, success has not been achieved with hybridisations between widely-different plants, but, in time, the technique may be practical to use for those which cannot naturally cross-pollinate. There would then be the possibility of bringing together in one hybrid several advantageous features (such as disease-resistance, nitrogen-fixing ability, rapid growth) normally found separately in different species. The possibilities are increased by the fact that, at the naked-protoplast stage, there is the opportunity to insert DNA, plasmids or phages.

Ethical considerations

Some people who are qualified to express an opinion on genetic-engineering techniques consider that there are biological dangers that should be taken into account. Another kind of disquiet concerns the problem of whether we have any right to 'meddle' with the hereditary endowment of living organisms in this way. Since our understanding of how living creatures evolve is so incomplete, moreover, critics argue that we are likely to make decisions whose long-term consequences we cannot possibly foresee. In any case, human-guided evolution is likely to be short-sighted, since our vision of what characteristics are 'good' is necessarily limited.

In spite of all this, human-beings *will* determine the direction of their own evolution. If conscious genetic choices do not do so, economic, religious and political choices certainly will. It is essential to make the best-informed decisions that we can at any time. The biological concept of polymorphism perhaps points to sound general principles: it is unwise to select only in one direction. We shall be safer in the long run if we do not limit ourselves to manipulating human evolution towards one ideal, however apparently laudable, but maintain a richly-diverse gene-pool with which to meet whatever the future brings.

QUESTIONS

1   In what ways may genetic engineering prove to be (a) beneficial and (b) harmful to living organisms? How may the risks be minimised?

Suggestions for Further Reading

TEXTS RELEVANT TO SPECIFIC CHAPTERS

Chapter 1
J. Huxley and H. B. D. Kettlewell, *Charles Darwin and his world* (Thames and Hudson, 1965).
D. Lack, *Darwin's finches* (Harper's Torchbook, 1947).

Chapter 3
K. R. Lewis and B. John, *The matter of Mendelian heredity* (J. and A. Churchill, 1965).
J. McLeish and B. Snoad, *Looking at chromosomes*, 2nd ed. (Macmillan, 1972).
M. J. D. White, *The chromosomes*, 6th ed. (Chapman and Hall, 1973).

Chapter 4
J. D. Watson, *The double helix* (Weidenfeld and Nicolson, 1968).

Chapter 7
A. J. Cain, *Animal species and their evolution* (Hutchinson, 1954).
W. H. Dowdeswell, *The mechanism of evolution* (Heinemann, 1975).
E. B. Ford, *Ecological genetics*, 4th ed. (Chapman and Hall, 1975).

Chapter 8
C. R. Austin and R. V. Short, *Reproduction in mammals* (Cambridge University Press, 1972). (Five books in a series.)
R. J. Demarest and J. J. Sciarra, *Conception, birth and contraception* (Hodder and Stoughton, 1969).

Chapter 9
C. R. Austin and R. V. Short, *Reproduction in mammals* (Cambridge University Press, 1972). (Five books in a series.)
B. I. Balinsky, *An introduction to embryology*, 4th ed. (W. B. Saunders and Co., 1975).
J. D. Ebert and I. M. Sussex, *Interacting systems in development*, 2nd ed. (Rinehart and Winston, 1970).
W. H. Freeman and B. Bracegirdle, *An atlas of embryology*, 2nd ed. (Heinemann, 1967, reprinted 1975).
A. R. Gemmell, *Developmental plant anatomy*, Studies in Biology series no. 15, (Edward Arnold, 1969).

D. R. Newth, *Animal growth and development*, Studies in Biology series no. 24, (Edward Arnold, 1970).
T. A. Steeves and I. M. Sussex, *Patterns in plant development* (Prentice-Hall, 1972).

Chapter 11
A. G. Cains-Smith, *The life puzzle* (Oliver and Boyd, 1971).
C. Ponnamperuma, *The origins of life* (Thames and Hudson, 1972).

Chapter 12
S. Cole, *The Neolithic revolution*, 5th ed. (Trustees of the British Museum, 1970).
J. Van Lawick-Goodall, *In the shadow of man* (Fontana/Collins, 1974).
D. Pilbeam, *The evolution of man* (Thames and Hudson, 1970).

GENERAL TEXTS

Oxford Biology Readers ed. J. J. Head (Oxford University Press):
    J. L. Jinks, *Cytoplasmic inheritance*.
    E. G. Jordan, *The nucleolus*.
    D. H. Northcote, *Differentiation in higher plants*.
    A. A. Travers, *Transcription of DNA*.
    L. Wolpert, *The development of pattern and form in animals*.

The molecular basis of life, ed. R. H. Haynes and P. C. Hanawalt (W. H. Freeman, 1968). A collection of relevant Scientific American articles.
Readings in genetics and evolution, ed. J. J. Head (Oxford University Press, 1973). A collection of Oxford Biology Readers.
P. Hutchinson, *Evolution explained* (David and Charles, 1974).
F. B. Salisbury and R. V. Parke, *Vascular plants: form and function*, 2nd ed., (Fundamentals of Botany series, 1973). Chapters 4–9, 13, 14 and 16–18 are relevant to chapters 5, 8, 9 and 10 of this book.
Facets of genetics, ed. A. M. Srb, R. D. Owen and R. S. Edgar (W. H. Freeman, 1970). A collection of Scientific American articles.

Index

Page numbers in bold type indicate major references and the pages where terms are first defined (i.e. where they appear in italic type in the text). Page numbers in italic type refer to illustrations.